A PARAMEDIC'S TALES

GRAEME TAYLOR

A PARAMEDIC'S TALES

• • • *Hilarious, Horrible and Heartwarming True Stories* • • •

HARBOUR PUBLISHING

1 2 3 4 5 — 24 23 22 21 20

Harbour Publishing Co. Ltd.
P.O. Box 219, Madeira Park, BC, V0N 2H0
www.harbourpublishing.com

Edited by Betty Keller
Cover design by Kim LaFave
Text design by Carleton Wilson
Printed and bound in Canada
Printed on 100 percent recycled paper

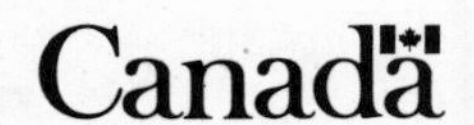

Harbour Publishing acknowledges the support of the Canada Council for the Arts, the Government of Canada, and the Province of British Columbia through the BC Arts Council.

Library and Archives Canada Cataloguing in Publication

Title: A paramedic's tales : hilarious, horrible and heartwarming true stories / Graeme Taylor.
Names: Taylor, Graeme, author.
Identifiers: Canadiana (print) 20190232137 | Canadiana (ebook) 20190232218 | ISBN 9781550179026 (softcover) | ISBN 9781550179033 (HTML)
Subjects: LCSH: Taylor, Graeme. | LCSH: Emergency medical technicians—British Columbia—Biography. | LCSH: Emergency medicine—British Columbia. | LCSH: Ambulance service—British Columbia. | LCGFT: Autobiographies.
Classification: LCC RA645.7.C3 T39 2020 | DDC 616.02/5092–dc23

For Lars and Zara

Contents

Some BC Ambulance Service Terms and Radio Codes

10-4 – message understood

10-7 – off air/at scene (crew leaving the ambulance)

10-8 – on air/on the way (crew in the ambulance)

ALS – Advanced Life Support

Ambu bag and mask – a manual resuscitator used to help patients breathe

BCAS – British Columbia Ambulance Service

Block – a group of shifts (e.g., two day shifts followed by two night shifts)

BLS – Basic Life Support

Call – request for an ambulance (e.g., "the car is out on a call")

Car – ambulance

Code 2 – routine (normal speed)

Code 3 – emergency (lights and sirens)

Code 4 – dead

Code 5 – police

Code 6 – firefighters

Code X – ambulance not used (e.g., "We're Code X" or "We're ANU")

Crew – ambulance crew (normally a driver and an attendant)

EMA – emergency medical attendant (a paramedic)

ERP – emergency room physician

Intubating – placing a flexible tube into the trachea to maintain an open airway

IV – intravenous (e.g., "Start an IV line")

Jump kit – a large bag with emergency medical equipment and supplies

Laryngoscope – a metal device designed to facilitate visualization of the throat

MVA – motor vehicle accident

OR – operating room

Oral airway – a rigid plastic tube used to maintain or open a patient's airway

Unit chief – ambulance station chief

Zero – break

Author's Note

While these stories are based on true events, many names, places and other details have been changed to protect the privacy of patients, their families and medical staff. And no, I didn't write about your aunt—lots of people get stuck in the toilet! Don't bother taking me to court—I'll deny everything.

I don't want my readers to think that emergency medicine is one long series of screw-ups—ambulances sliding into ditches, stretchers careening down cliffs, paramedics discussing sports with dead patients. I write about such weird and wonderful events because most people wouldn't read a book describing typical ambulance calls: that is, the ones where the patient has a heart attack, is treated with medications A, B and C, stabilized and transported to hospital. But ambulance work also proves Murphy's Law: sooner or later anything that can go wrong will go wrong ... and make a great story!

Acknowledgements

These stories combine my own experiences (I kept detailed notes throughout my career), with stories told to me by co-workers. My thanks to all the ambulance paramedics and dispatchers who contributed to this collection. I have retold them in my own way, and I bear complete responsibility for the depictions of people and events.

A Paramedic's Tales would also never have been written without my wife Ferie's encouragement (and prodding), and without the support of Harbour Publishing and its excellent staff. In particular, I am grateful for the patient editorial guidance of Betty Keller and Arlene Prunkl, who had to put up with an author who kept adding new material long after the maximum word total had been exceeded and final deadlines passed.

Above all I would like to express my gratitude to the British Columbia Ambulance Service for granting me the opportunity to become a paramedic, and to the many people I had the privilege to work with. I am especially indebted to Michael Wheatley and Mike Havard, who kept me going with their friendship and good humour during my final years working on ambulances, when my back was giving out and my nerves beginning to fray.

Introduction

It happened almost every time I met new people. They would start with, "And what do you do for a living?"

"I'm a paramedic."

"A paramedic? You mean you drive an ambulance?" they would ask, getting strange looks on their faces. I was never sure whether it was awe or disgust.

"Yeah, I drive an ambulance."

"I wouldn't want your job! How do you do it?"

"Well," I'd answer, "it sure beats working nine to five."

"But you must see a lot of terrible things!"

"Yes." I hoped they would lose interest if I kept my answers really short. But they rarely lost interest.

"How can you stand the bleeding and all those dead people?" they would persist.

I would try being honest. "Actually, I'm a vegetarian," I would say. "I don't like blood or dead things. That's why I cover wounds with bandages if people are alive or stuff them in body bags if they aren't."

Most people changed the topic after that. In fact, most people stopped talking to me altogether. But some brave ones just wouldn't quit. "You must have seen some pretty strange things!" they would say, eager to hear something really gruesome and depressing.

"All right," I would reply, giving in. "Let me tell you about the time my partner and I were carrying an old guy out of a swamp and his dentures fell out ..." And I would start on a story that I found absolutely hilarious, never knowing whether my audience would crack up or turn away, appalled by my insensitivity to human suffering. The thing is, you won't survive long in emergency medicine if you don't have a black sense of humour because enough tragedies occur every day to keep all the angels weeping.

But that wasn't all I saw on this job. Paramedics are exposed to a vast spectrum of human experience, often in the most surprising and delightful ways. And the best part of working in this profession is that it is one endless adventure. You never know what will happen next—where you will go, who you will meet or what you will be asked to do. You show up for work, and life takes you from there, usually at high speed.

Part 1: How I Got Hired

As a teenager, I worked in Ottawa as a medical orderly, in Toronto as a psychiatric attendant and then in Edmonton as an operating room orderly. This got me used to seeing what people look like on the inside as well as the outside and got me interested in the fascinating world of medicine. Of course, not everyone reacts this way to such experiences. Once I was in a hospital elevator on my way to the morgue. A group of visitors got on and one guy asked, "What's in that big box?"

"You don't want to know," I replied.

He persisted. "Yes, we do! What's in it?"

"A leg," I said.

They laughed and a woman said, "We don't believe you. Show us!"

So I did.

They all rushed out at the next floor.

Supernatural British Columbia

I first saw British Columbia in the fall. I arrived from Toronto with my friend Liz, a professional photographer, who wanted to take a break before she started a year-long assignment in South America. We'd heard that hippies were squatting in abandoned fishermen's shacks on the north end of Galiano Island, so we went there and appropriated an old moss-coated log cabin for our vacation.

The first morning dawned fresh and bright, with sunlight streaming down between tall fir trees and dappling the high ferns surrounding our one-room house. Beyond the trees the light glinted off white surf and a gently heaving sea. The previous night we had sat around a bonfire in the sand, surrounded by huge pieces of driftwood, watching sparks winding up into blackness while we chatted with the mellow, long-haired family that lived next door. Then we had lain awake freezing as a damp ocean wind blew in through gaps in the old log walls.

But now the cold night was gone, transformed into one of those perfect fall days where the crisp air is warmed by a brilliant sun shining out of an almost cloudless sky. The crystal salt air was invigorating, and Liz decided to hike along the trail that led to a nearby store and pick up some food, while I felt inspired enough to wander along the shoreline, gazing through the clear water at purple starfish and orange crabs. But then I spotted a small wooden dinghy that someone had left behind, complete with oars, and my horizons expanded. I would row down to the store, surprise Liz and bring her and the groceries back to the cabin.

I dragged the boat down to the water and pushed it out through the low breakers, jumping in at the last moment. I then rowed slowly away from the shore, past the long, trailing fingers of kelp, hearing nothing but the slap of the water on my oars, the occasional bark of a seal and the crying of seagulls floating on the light breeze. It was truly a perfect day in a perfect place, and I lay down in the softly rocking boat to watch the few spotlessly white clouds drifting across the sky.

Finally I headed toward the store, rowing into the pass that divides Galiano from Valdes Island to the north. I passed an Indigenous fisherman, who was hauling up dogfish. "Hi!" I called out cheerfully. "Glorious day, eh?" He glanced up at me briefly, said nothing and went back to his fishing.

The ocean was calm and smooth, and I rowed along easily, watching the black diving birds flying low over the water. A hundred metres to my left the shore went by slowly, the still green forest separated from the gently pulsing blue of the sea by sensuously twisted grey and brown rocks. A happiness filled me. I had grown up on rivers, and here

I was once again in that magical place where moving water meets a changing sky, except that on the ocean everything is much bigger and much more beautiful.

Suddenly I was conscious that the sea was growing rougher and filling with short, sharp waves. It was not particularly dangerous, but it was puzzling, as the same light breeze blew and a million fragments of sun sparkled reassuringly on the water. I looked around and saw that the shoreline was starting to pass by more quickly. What? I was in some sort of current, being pulled into the narrow channel between the two islands.

The waves were becoming steeper and starting to crest with angry whitecaps. I was beginning to feel alarmed, and I tried to carefully turn the dinghy toward the shore. But it was already too rough—I would capsize if I did anything but run before the waves down the channel. I tried rowing backwards, but it was useless. The current was too strong for me, and all I could do was sit facing forward, brace my legs against the violent jerking and try to keep the little boat from shipping too much water.

What a bizarre predicament! I had no idea what was happening. I decided to risk standing for an instant to see where the current was taking me. Up ahead was a stretch of wild water, and beyond it the unmistakable contours of an enormous whirlpool, eight to ten metres across. This wasn't possible! Whirlpools didn't exist in oceans, did they? I had once read an Edgar Allan Poe story about a ship that sank in a gigantic maelstrom, but I always thought he had an overactive imagination. But there it was—and here I was, being sucked toward an inevitable, swirling death.

I sat back in my bucking boat and looked around at the placid sky and peaceful forests. What a beautiful day to die! I guess death always catches you by surprise—one moment you are alive, and then your time is up. If I have to die, I might as well die on a perfect fall day like this. I felt strangely calm and rose again to have another look at my absurd doom, which had already grown closer and clearer.

Then I saw a large powerboat close by Valdes Island on my right, steadily pushing against the current. I waved frantically, and one of the fishermen on board waved back. In a few minutes they had pulled alongside and thrown me a line. I wrapped it around my wrist, and after another ten painful minutes my dinghy and I had been towed back out of the current and were lying behind the powerboat in calm water.

"What the hell were you doing in that riptide?" one of the men called to me as he reeled in his rope. "Quite a few big yachts and tugs have been sucked into that whirlpool, and it spit 'em out in little pieces."

"Sorry about that," I said. "This is my first time out here. I'm from Toronto."

"Toronto? That explains it." The men gunned their motorboat and tore off, poking each other. "He's from Toronto! Ha ha!"

I felt like a total moron. "Uh, thanks!" I yelled at the retreating froth.

I rowed back slowly over the sunny sea, a little suspicious that some other legendary nightmare might rise up to swallow me. I passed the Indigenous fisherman, still in the same spot catching dogfish. I called across to him, "I almost drowned in that whirlpool!" Again he looked up briefly, nodded and went back to work. Clearly, warning brainless hippies wasn't high on his priority list.

I pulled the boat back up the beach just as Liz was returning with a backpack full of groceries. "Liz," I called. "I just had an amazing experience out there!"

She turned on me angrily. "Why don't you try helping for a change? Here I am, hot and tired from carrying all this stuff back from the store, and all you can think about is telling me another one of your stupid stories!"

No Point Panicking

After that adventure I couldn't go back to a routine life in Toronto, so I moved to Vancouver Island and found a job in a sawmill pulling lumber off the green chain. It was great exercise—like working out in a gym all day—but it soon became boring, and five minutes after I began work in the morning I'd start counting the minutes to the coffee break.

Fortunately, every other shift a sliver would stab right through my leather gloves and give me an excuse to leave the chain to get it pulled out. What a pleasure—paid time off! I'd walk slowly up to the shack with the red cross on the door. There the first-aider would put down his newspaper, get out a scalpel and a pair of tweezers, pull out the sliver and bandage me up, and fifteen minutes later I would be back working on the chain and he would be back reading his paper.

One day I said, "Your job sure beats stacking boards all day. How did you get it?"

"First you need a Worker's Compensation Board first-aid ticket, and then you have to have at least fifteen years' seniority." And he laughed. "So I'm not worried about you stealing my job. But we do need someone to cover my holidays—you'd only need a 'C' ticket for that."

Right away I registered for a first-aid course, and that summer I spent a glorious two weeks with my feet up on his desk, my reading only occasionally interrupted by someone needing a foreign object removed from an eye or limb. Since I didn't want to spend my life humping wood, when it was time to go back on the chain, I quit and enrolled in a graphic arts course in Vancouver. Because my application stated that I had WCB certification, I was designated the class first-aid attendant.

In one of our printing classes, my assignment was to run an old Heidelberg platen press—a beautiful piece of machinery used to print and number tickets. A few metres in front of my machine was a much larger full-colour press manned by two of my classmates. The one directly in my line of sight with his back to me was operating the paper feeder. But I was mostly oblivious to my surroundings because I was focused on keeping the ink flowing smoothly and mesmerized by the steady rhythm of my press as it picked up the tickets one by one, numbered them and stacked them in a pile—*kerthunk, kerthunk*. Suddenly I realized that the entire class was crowding around the next press, and I turned off my machine, wiped my hands and sauntered over. The feeder operator was now naked from the waist up. I could see the remains of his shirt in the rollers of his printing press. But although he looked badly shaken, he didn't seem to be injured.

"Are you okay?" I asked.

"Uh ... yuh ... yeah," he stuttered. "I think so."

"Well then," I said, "why don't you find something warm to wear, sit down and have some coffee." And since there didn't seem to be anything else to do, I went back to my press.

A short time later another classmate came up to me. "Wow," he said admiringly, "you're really cool under pressure. I saw the whole thing—you didn't even blink when the gears ripped off Fred's shirt. But I guess you could see he was okay."

"Sure," I replied. "No point panicking over nothing." Of course, everyone in the school talked about the incident, and by the end of the week my reputation as an experienced medic was firmly established.

Inevitably there was a real emergency. A middle-aged man collapsed outside the college, and they sent for me. I ran outside and looked at the body sprawled in the middle of the street. Umm, was he alive or dead? I felt for a breath and then a pulse. Yes, he was alive. But what was wrong with him? Had he been hit by a car? Had he had a heart attack, a stroke? Was it a diabetic coma, an internal bleed, an overdose? I was thinking about what I should do next when an ambulance pulled up, lights flashing. Two paramedics stepped out, opened the back doors and in one easy movement slid out the stretcher and lowered it to the ground. One of them waved me away. He rapidly assessed the patient's vitals—including his level of consciousness and pupils—checked him from head to toe for injuries, pulled an airway out of his bag and slipped it into the patient's mouth. He and his partner then lifted the man onto the stretcher and into the ambulance. The driver quickly closed the back doors and jumped into the front seat. In a minute the ambulance was gone, the sound of its siren receding in the distance. The onlookers drifted away and traffic returned to normal.

That was impressive, I thought. They actually know what they're doing. I should learn to do that!

Meat Wagon

I finished my graphic arts course, but my heart wasn't in it. The printing solvents gave me headaches and I hated typesetting. So I upgraded my first-aid qualifications to a "B" ticket, got a Class 4 driver's licence (to allow me to drive ambulances), and filed an application with the newly established Provincial Ambulance Service. But they weren't hiring, so I went to work as a full-time first-aid attendant in northern Vancouver Island logging camps.

You'd think it would have been the perfect job for a guy who likes to read. I sat in a comfortable office while hundreds of fallers, chokermen, mechanics and heavy equipment operators sweated and slipped on the steep hillsides. Although I had to deal with the occasional ugly chainsaw cut or broken leg, I was lucky that the really bad accidents—the operator of a yarder smashed by a runaway log, a manager decapitated by a helicopter's rear rotor—always took place on someone else's shift. Nevertheless, I soon discovered that life in isolated camps is not good for the health.

At three one morning I was awakened by a loud hammering on my bunkhouse door. "Hey, Bandaid, get out here and patch me up!"

"Okay, okay!" I yelled back. "Don't smash down the door—I'm coming!"

Carl, a tough ex-con and faller, was standing on the step, dripping blood from deep lacerations to his forehead and lower lip. He grabbed me by the arm and dragged me across the road to the first-aid room. "Patch me up!" he demanded, pushing me inside. A massive slab of muscle, he was intimidating at the best of times, but now he was also drunk and angry.

"What happened?" I said as I started cleaning his lower lip.

"I beat Bill at poker, and he hit me from behind and put the boots to me. I need you to patch me up so I can get my gun and kill the bastard. What the hell's taking you so long? Hurry up!"

It wasn't easy to clean and bandage Carl's wounds. His lower lip was almost severed and he was flailing his arms furiously so that whenever I got close, he would knock me away. "Look, Carl," I said, trying to calm him down, "you're too drunk to shoot straight. Why don't you get some sleep—you can kill Bill in the morning."

"Yeah, maybe," he mumbled. "I'll kill him tomorrow."

First thing in the morning I went to see Arnie, the camp bull [top boss]. When I knocked on his office door, he barked, "What the hell do you want? It'd better be important 'cause I'm busy."

"I'm sorry to interrupt you, boss, but Carl and Bill got into a fight last night, and Carl got beat up pretty badly. He wants revenge and says he's going to kill Bill today. You have to stop him."

Arnie stared at me in disbelief. "Why? What do you think this is, a kindergarten? Loggers like to drink and they like to fight. We pay you to drive the meat wagon, not play social worker. If someone gets killed, it's your job to get rid of the body. Now get out and don't bother me again."

I worked in logging camps for two years, which was long enough to fall in love with the tall trees and soft mists of the North Island. It was also long enough to be away from home and destroy a budding relationship. I had made friends among these alpha males, and part of me was going to miss living in a tightly knit wolf pack. But every day these hard-working men were clear-cutting thousand-year-old forests—actions that future generations would compare with blowing up museums filled with priceless antiquities. And their guilt was my guilt: those huge tree trunks filling the logging trucks were paying my wages.

So when the camp broke up for the winter shutdown, I packed my gear in my car and took a last look at the snow-covered hills surrounding the camp. The cooks had already gone and the dozens of cats that lived on scraps beneath the cookhouse were getting hungry. Some were starting to wander around the deserted buildings, and I watched while eagles swept down and carried them away.

A Government Job at Last

I took a long holiday in Europe, returning to the West Coast in the summer with a red-haired fiancée, an empty bank account and a problem: if I was going to get married and settle down, I needed to get a steady job. So as soon as I had unpacked, I dropped in at the Ambulance Service's regional office in Victoria. By now the secretary knew me by name.

I greeted her and asked, "Will you be hiring soon?"

"Guess what?" she said. "They're conducting interviews right now. I can fit you in in fifteen minutes."

What a stroke of luck! By now I had a fair bit of medical experience and a WCB "A" ticket, and a week later I received a letter telling me that I had made the shortlist. All I had to do was pass a physical competition. No problem! Even after two years of sitting around in a first-aid shack I could still do 50 chin-ups and 150 push-ups, so passing a physical test should be easy. And then I would be set for life with a government job, decent salary, vacations and a pension.

The following Tuesday I joined a small crowd of would-be paramedics at a playing field. To my shock forty applicants had been shortlisted—and we were there to compete for just four jobs! I broke into a sudden sweat; I should never have gone partying last night!

The competition was more marine corps than ambulance. For four hours we were timed and tested on everything from carrying loaded stretchers up flights of stairs to climbing over obstacles and running a hundred-yard dash and then a mile. When they added up the points, I came in fourth—I had just made the cut. But here again I had been lucky. One of the applicants was a tall, athletic-looking guy—a St. John's first-aid instructor and football player. He had been training every day for a month for this competition, and I'm sure he would have beaten me if he hadn't managed to sprain his ankle the day before. As it was, he showed up on crutches, and during the running events gamely hobbled after us around the track. Sometimes it pays to party!

Part II: Vancouver and the Lower Mainland

Learning the Ropes

After I was hired, Karin and I married and moved to Vancouver, where I received basic training in emergency medicine and ambulance work at the Justice Institute of British Columbia. As soon as I graduated, I started working in the Vancouver area.

It can get really busy for paramedics in the Lower Mainland. Ambulances can leave their stations at the start of a ten-hour day shift or fourteen-hour night shift and not return for twelve, fourteen or even sixteen hours. Some days and nights the calls would come back to back without leaving time for a coffee break, let alone time to return to the station to restock supplies. I quickly learned not only to pack a lunch but also to show up thirty minutes early to thoroughly check the ambulance, ensuring it was clean and safe with lights, sirens, tires and other mechanical parts in order and that it was fully stocked with equipment and supplies—stretchers, backboards, gases, defibrillator, drugs, splints, dressings, IV catheters and tubing, bedding and a hundred other items from restraints to spare gaskets.

MY FIRST DELIVERY

At the end of our paramedic training period, each trainee was assigned to a regular ambulance as a third crew member, and at 5:00 p.m.—the start of a fourteen-hour night shift—I waited tensely for my first call.

The hotline rang two minutes later. "Code 3 for a woman in labour."

One of the paramedics told me to climb in the back. "You're attending. As soon as we get to the call, go ahead and start assessing the patient—we'll follow with the stretcher."

As we raced through the streets, sirens wailing, I tried to remember how to deliver a baby and what to do in the case of complications: if it was a breech birth, if the cord was wrapped around the baby's neck, and so on. As I added an infant Ambu bag to my jump kit and checked and rechecked the contents of the maternity kit, I could feel sweat running down the back of my shirt.

The driver stopped the ambulance at a house with peeling white paint, and I threw open the side door, leaped out and ran up the path. Two teenaged girls were walking down the front steps. I paused and asked, "Is someone here having a baby?"

"Uh, yeah ... I guess so," one replied. They kept on walking.

I pushed open the front door and found myself in a short hallway. To the left was a living room where three young men lay on couches. They looked asleep. "Hello!" I called out. "Hello?" No answer. I went farther into the house. "Hello! Is anyone here?"

A female voice answered. "Who are you?" She was behind a locked door.

"I'm a paramedic. Are you having a baby?"

"Yes."

"Can you open the door so I can look at you?"

"No—I'm on the toilet."

"Stop! Don't have your baby in the toilet! Let me get you on a stretcher."

"I'm not having the baby now. It's not due for another six months."

Confused, I turned around. What was going on? I walked back down the hallway. The three guys were still passed out in the living room. But something looked wrong. I went over to take a closer look. The first one was barely breathing. Wait ... they were all barely breathing! I pulled an airway out of my belt, put it in the nearest man's mouth and started to ventilate him. Just then my partners showed up with the gurney.

"I think we've got three overdoses!" I yelled. They swung into action. In a few minutes we had two patients loaded into the ambulance and were on our way to the nearest hospital. One of the experienced paramedics was attending—I had suddenly been promoted to driver. The other paramedic stayed behind with the third OD and waited for a backup car.

All three druggies survived.

On reflection I figured the girls had realized that the three guys needed help, and they called for an ambulance. Because they didn't want to attract the attention of the police, they told the dispatcher that a woman was giving birth.

I never saw the pregnant woman. Half an hour later we had cleared emergency and were off on the second call of a long and busy shift. But I had learned an important lesson: don't make assumptions. You won't know what's going on until you arrive at the scene, so you might as well relax and enjoy the ride. You'll find out what's happening soon enough.

THE DAY I ALMOST KILLED TERRY FOX

As soon as I was considered ready to work on my own, I was placed on the Lower Mainland spare board, filling in for crew members who were either sick or on holidays. On almost every shift I was sent to a new station located somewhere I had never been and partnered with someone I had never met. It was pretty stressful. This was long before GPS devices made driving easy, and I either drove nervously down strange roads, searching in the dark for accidents, houses or hospitals, or sat in the back of the ambulance in a cold sweat, hoping that I was giving the right treatment to a seriously injured patient.

Finding your way around the Lower Mainland can be difficult, especially if you're a newcomer. Few of the streets run straight; they bend and dead end in complex patterns dictated by the mountains, the rivers and the sea.

One day I received a call from a supervisor asking me to work the next Sunday at the station in Coquitlam. I told him that previously I had worked only one shift out there as an attendant and didn't know the area at all. "I'll only go out there if you can pair me with an experienced driver," I said.

"No problem," came the reply. "I'll put you with one of the regular crew members from the area."

I showed up at the Coquitlam station early Sunday morning and shook hands with Nick, my new partner. "I guess you'll be driving," I said.

Nick looked shocked. "What? I just transferred from the Okanagan yesterday. I've never driven anywhere in the Lower Mainland."

"You're kidding," I said, aghast. "You mean I'm the one with the most experience around here?" We were the only ambulance covering Coquitlam, Port Moody and Port Coquitlam, and I had never set foot in the last two! "Well, you'll have to drive anyway," I said.

My new partner shook his head. "I don't even know how to find the main road to Vancouver."

A sudden thought struck me. "This is terrible! Terry Fox's home is in Port Coquitlam, and he's only expected to live a few more days. He's the best-loved person in the whole country, and if we get a call for him, we'd better be able to find his house."

Nick paled. Terry Fox had run halfway across the country on an artificial leg to raise funds for cancer research. "You're right," he said. "Let's get out of here and drive around until we get our bearings."

I sighed. "Okay, I'll drive. Seems I have no choice." Very quickly we checked out the ambulance, and I drove it out of the station and headed down some side streets at random.

"You get out the map and navigate," I said. "We'll find our way back down to the centre of town and then head over to Port Coquitlam."

Nick looked at the nearest street sign, then searched for it in the map index. "It isn't listed!" he exclaimed. "We must be in a new subdivision."

"Okay, we'll drive out of it to someplace that's on the map," I said. "Now which way did we come?"

"I don't know," my navigator replied miserably. "I'm totally lost."

"Wonderful," I said. "Just wonderful. Now we're going to get the call for Terry Fox, and we don't even know where *we* are, let alone where he is." My hands were clammy on the wheel. "Let's just get back to the station and sit tight."

After a few more twists and turns we made it back out to a main road and found our way to the station. We sat waiting nervously for the first call. It came soon enough: take a patient with terminal cancer from Port Coquitlam to the Royal Columbian Hospital in New Westminster.

I tried to be calm as I studied the map. "Here it is," I said, pointing. "Now guide me in carefully. We can't make any mistakes. There could be TV crews waiting at the house. If we get lost getting there, show up half an hour late, and then get lost going to the hospital with six cars of journalists tailing us, we're not only going to lose our jobs, but the public will despise us so much we'll have to leave the country!"

But miraculously, we drove straight to Port Coquitlam and found the house without a problem. We walked inside, but instead of Terry Fox, the patient was a man in his sixties with cancer throughout his abdomen. He looked like he was on his last legs, so we lifted him very gently and took him out to the ambulance.

"I'll follow you to the hospital," said his son.

"Uh … perhaps you'd like to go ahead and find a place to park," I offered, hoping I could follow him. But he stayed behind to lock up.

I found my way to the Lougheed Highway with no trouble, and from there I could see the Port Mann Bridge ahead and to my left. My partner in the back of the ambulance suddenly leaned forward and spoke quietly into my ear: "Maybe you should speed up a little. He's fading out. His respirations are down to ten per minute. I'm going to start bagging [ventilating] him."

"Okay," I replied. "We should be there soon. I think I can see the hospital." I put on the lights and siren and turned off the first exit I came to. The next thing I knew I was heading across the Port Mann Bridge—not where we wanted to be!—and a high divider prevented me from turning around.

What should I do? In the distance on my right I could see the Pattullo Bridge arching across from Surrey into New Westminster. Should I try finding it? But what if I got hopelessly lost in Surrey? I glanced in the mirror to see whether the son was chasing me around, but I couldn't remember what his car looked like.

Thankfully, on the other side of the bridge the divider ended, and I made a quick U-turn. The exit I chose this time took us directly to the Royal Columbian Hospital. The son was nowhere to be seen—he was probably parking his car—and we hurriedly pushed the stretcher into the emergency room and lifted the semi-conscious man into a bed.

As soon as we put him down, he stopped breathing. We bagged him again and once more his chest began to slowly move. At this point we gave a brief report to the nurse and made our escape.

"Thank God that wasn't Terry Fox!" the two of us exclaimed simultaneously as we walked out of emergency. For the rest of the shift we waited tensely for another call to Port Coquitlam, but it didn't come. Terry Fox was taken into hospital for the last time a few days later, fortunately by a crew who knew where they were going.

THE WRONG CALL

One November morning I was sent to a downtown station for a twelve-hour shift. As it was to start at 7:00, I arrived at 6:30 and immediately went to the garage where three ambulances were parked. I was told that I would be riding in the second one, and I had just climbed into the cab to test the lights when a burly paramedic walked into the bay and headed for the first ambulance. He saw me and called, "You must be my attendant. Jump in—we're Code 3 for a stroke!"

I grabbed my lunch and climbed into the ambulance he was pointing to. He said, "I'm Ben," and took off at high speed, lights flashing and sirens wailing. In a few minutes we arrived at a tall house. As we lifted out the stretcher, we could see a fire truck turning into the street.

An older woman opened the front door. "I'm his landlady. Follow me." We carried the heavy stretcher up three flights of stairs to a wide hallway. At the end a metal stairway spiralled up to a room in the attic. The woman pointed. "He's up there."

"We'll never get this stretcher up those stairs," my partner said. "They're way too narrow. Let's see what we've got."

We hurried up the spiral stairs with the jump kit and oxygen bottle. A big, balding man lay on a single bed with one arm over the side. His jaw was slack and his breaths made rasping sounds. I flashed a light in his eyes: unequal pupils. I yelled in his ear and squeezed his traps: no response. His systolic blood pressure was over 200. "Looks like a bad CVA," I said, inserting an airway into his mouth and strapping on an oxygen mask.

"We'll need to whistle him out of here," Ben said as he connected O_2 tubing and turned the regulator to high flow. "But first we have to get him down those stairs, and that won't be easy—he must weigh over three hundred pounds. I'll send a fireman for the scoop stretcher."

A few minutes later we had snapped the two halves of the scoop together under the patient and strapped him tightly to its light metal frame. I lifted the head and Ben lifted the foot and we carried our patient to the top of the spiral staircase. There Ben paused. "Damn! We'll have to raise him high enough to clear the handrails. You lift him to shoulder height," he told me, "and I'll extend my arms straight above my head to keep him level." Then he turned to the firemen. "One of you guys get behind me to keep me from falling backwards."

Ben and I hoisted the sagging stretcher up in the air and stepped slowly down and around the staircase. One fireman leaned against Ben's back while another followed closely behind me with the oxygen tank.

I didn't know whether I would make it all the way. Our patient was huge—my muscles felt like they were ripping—and the pain was excruciating. But we managed to get to the hallway, lower the scoop on top of our cot and tie down the oxygen tank. Now things became easier. The firemen helped us carry our patient down the three flights to the street and lift him into the ambulance. Ten minutes later a triage nurse was leading us to an empty emergency bed. There another crew helped us transfer the patient and retrieve our scoop.

I put the equipment back in the ambulance while Ben called dispatch. He came outside, shaking his head. "I'm supposed to take you back to the station. Apparently you're on the wrong car."

A tall blond paramedic was waiting for me in the garage. "Where the hell have you been? They've been holding a call for us for ten minutes!" As I climbed back into the second ambulance, I looked at my watch: ten minutes past seven.

My ten-hour shift was just starting and I was already exhausted. My uniform was soaked in sweat and every muscle ached. The worst part was that I had wrecked myself for nothing. Because I wasn't supposed to have been on that call, I couldn't put in for overtime. I had been working for free!

Desecrating the Dead

When the British Columbia Ambulance Service was formed in 1974, it amalgamated all the existing private and municipal ambulance services. Although many of the people working for these services had little medical training—some private ambulances were essentially horizontal taxis—they were kept on staff and then partnered with new, better-trained employees.

So as soon as Sean, a new hire, finished his course in emergency medicine, he was partnered with Jake, a big, friendly man who had worked for a decade for a private ambulance service. Jake knew Vancouver like the palm of his hand, but he had no idea how many bones were in his wrist. On their second shift they were sent for a collapse. Arriving at the house, they were ushered into the living room where a man was lying on a couch, surrounded by family members.

"My dad isn't breathing," a young woman said. "I think he's dead!"

Sean bent over to check the man's breathing and pulse, then pushed the coffee table out of the way, pulled the man to the floor and started pumping his chest.

Jake was horrified. His new partner had gone nuts! He bent close to Sean's ear. "Sean, what are you doing? You're desecrating a corpse! Stop and show some respect!"

FILL YOUR BOOTS

Fortunately, in my early days on the job I was usually teamed up with experienced paramedics, most of whom were sympathetic and helpful, and I learned a lot in a hurry. With time, the job became less frightening and more fun. I also started to learn some paramedic slang. We worked closely with the fire department and the police, and some of the Vancouver crews had nicknamed the firefighters Cylons (because they dress like aliens and work in squads) and the cops Klingons (because they grab people). Sometimes I wondered what they called us. I never found out, although one wet, cold November evening I discovered what paramedics *should* be called.

We had transferred a patient to a top neurological facility and were about to clear the hospital when we were asked to help another ambulance crew move their patient, a twenty-one-year-old motorcyclist who had lost control and skidded into the side of a truck. On impact he had broken his neck, badly injuring his spinal cord. In that instant he had become a quadriplegic; still fully conscious but with no sensation or motor function from the neck down.

The ambulance crew had immobilized him on a backboard to ensure that no further damage was done during transport. Now the challenge was to get him into a hospital bed so he could be X-rayed and sent to the operating room. This meant carefully removing the neck brace, sandbags and straps holding him on the backboard and cutting off his clothes—leather jacket, leather pants, shirt, underwear and socks.

In order to avoid moving the patient's neck and body, two paramedics put their hands beneath his head and chest while another supported his legs. I stood between them with my right hand supporting his lower back and my left hand supporting his upper thighs. We lifted the young man to shoulder height and stood rock still while three nurses slowly cut off his clothing. All of us understood the need to avoid jarring the patient's injuries—the terrified motorcyclist didn't even move his lips.

The nurses worked cautiously and the minutes passed slowly. Then something warm and wet trickled onto my left hand and down my jacket sleeve. Oh, great—the man's bladder had let go! I willed myself not to move. The flow increased, running down my shirt and pants until my left boot was full of urine.

At last the patient was naked and we were able to lower him onto an X-ray transparent backboard on a hospital bed. We helped the nurses put on a new neck brace, surround him with sandbags and cover him with blankets. Then they wheeled him down a long corridor into the bowels of the hospital.

I called dispatch, booked off and drove home to shower, change and wash my clothes. But my discomfort was nothing compared to the tragedy I had just seen: one minute that young man had been happy and full of hope; the next minute he was in hell, unable to even scratch his nose. He was a person without a future. As I stood under the hot water, I thought how lucky I was to be healthy and have an interesting, worthwhile job.

And then the answer came to me: if firefighters are Cylons and police are Klingons, paramedics are Peeons!

An Impractical Joke

Oscar was hired by the Ambulance Service around the same time I was, and he became one of my classmates at the Justice Institute's Paramedic Academy. One day at lunch I asked him whether he'd had any previous medical experience. "No," he said, laughing. "Unless prepping bodies counts as medicine. Before this I was a mortician."

This surprised me as Oscar was a genial, garrulous fellow. I said, "Embalming corpses isn't my idea of fun. I haven't run into many dead people who can keep up a good conversation."

"Actually," he said, "the bodies are only a small part of the job. People who work in funeral homes have to be good at dealing with people because they spend most of their time organizing funerals or helping relatives fill out all the paperwork that goes with a death. My former boss, Mr. Brown, is exceptional at this. He's the perfect funeral director, always serious and sympathetic. In fact, he's too serious: he almost never cracks a smile." Oscar then told me he once devised a practical joke to help his boss loosen up: a real killer of a prank.

Mr. Brown always opened the funeral home at 7:00 a.m. sharp. He would then go straight to the mortuary room to make sure that the right bodies had been properly prepared for the day's funerals. But on this day Oscar had arrived early and gone into the cold mortuary room. He'd climbed onto an empty trolley, covered himself from head to toe with a white sheet—and waited.

Right on time Mr. Brown opened the door and turned on the light. Suddenly one of the bodies sat bolt upright and let out a loud scream. *"Ooohh!"*

Mr. Brown also screamed. He had been a mortician for thirty-five years and never before had anyone return from the dead. For a moment he thought he was having a heart attack. He almost collapsed from the shock.

And Oscar almost lost his job.

Emergency Driving

Paramedics usually don't like being called ambulance drivers because we don't just drive ambulances. We also diagnose and treat injuries and disease. In fact, driving is usually the easiest part of the job. The stressful part is calling the shots right in a critical situation and doing everything without error—for example, starting an IV in a dying patient when the blood pressure has collapsed and you can't find any veins. Of course, some paramedics will tell you that the really hard part of the job is forcing yourself to get up at four on a cold winter morning in order to carry some overweight patient complaining of sore muscles down a few flights of slippery steps.

However, this isn't to say that driving ambulances isn't an important part of the job. Or that it isn't sometimes stressful. Or that it isn't fun. The old line goes that 25 per cent of paramedics join up because they want to save lives; the remaining 75 per cent join because they want to run red lights. As for me, I like both aspects of the job: driving like the devil getting to a call, and then, having relieved all my aggression pushing other drivers off the road, I can act like a perfect angel with my patient.

But it's definitely not the right job for a nervous person. People can die if taxi drivers break the law, but in our case people can die if we *don't* break the law. We're allowed to do just about anything we like while on an emergency run—drive in the wrong lane, make U-turns in rush-hour traffic, park in the middle of the road and so on. But there is a catch: if we have an accident, we are responsible. So the art of ambulance driving is balancing the need to be quick with the need to be safe. The last thing I ever wanted to do was to come around a corner too fast and turn some kid playing road hockey into a hood ornament.

BRAKING PRACTICE

No one learns emergency driving overnight. When I first started, I was paired with a senior paramedic who was responsible for teaching me emergency driving. I was at the wheel, driving a wide-box ambulance along a fairly deserted stretch of highway in West Vancouver, when he suggested I practise emergency braking. It is important to know how rapidly an ambulance can stop without locking up the wheels and sliding.

I looked ahead. The road was clear. I glanced in my side mirrors. Nothing visible behind. I said, "Hang on!" to my partner and hit the brakes hard. Suddenly from behind us came the sound of screeching brakes and squealing tires, and a red Porsche careened out beside us and shot ahead. As the car overtook us, I had a brief glimpse of an ashen-faced driver.

I'm sure that's one driver who never tailgated an ambulance again!

HIT-AND-MISS STATISTICS

Managers in every ambulance service are always looking for ways to improve service while reducing costs. The easiest solution is to put the ambulances closer to where the calls will occur, which makes for quicker responses, wastes less gas, and shortens the time ambulances are tied up, enabling crews to get back in service faster and do more work.

Unfortunately illness and accidents don't take place according to strict timetables, and Murphy's Law guarantees that the semi will collide with the school bus in the middle of nowhere instead of somewhere convenient like right across the road from an emergency room.

Perhaps in the future we will be able to phone someone up and say, "Our calculations indicate that you will fall down your front steps and break your right leg next August 19 at 3:19 p.m. We will have an ambulance waiting for you and have already booked time in an operating room at the General Hospital. We have arranged for you to stay in room 422, bed C, for the following three nights. Enjoy your visit!" Unfortunately we still don't know how to connect psychics to dispatch computers, so our managers have to analyze statistics in order to calculate when and where ambulances are likely to be most needed.

Trying to make a few ambulances cover a large area is part of the art and stress of ambulance work, an art in which the managers handle the strategy (how many crews should be based where) and the dispatchers handle the tactics (where to position an ambulance at any particular time). Of course, sometimes managers can't resist getting involved in tactical planning.

One manager was poring over the stats when he noticed an unusual pattern emerging. An awful lot of motor vehicle accidents were occurring in the same area of East Vancouver. He pulled out all the relevant crew reports and started toting up the figures. Most of these accidents were taking place at the same intersection and on the same days of the week. In fact, the statistical probability was that on any one of those days at least one accident would occur between the hours of four and six in the afternoon. He called up the dispatch centre and requested that they experiment with putting an ambulance on standby at the intersection the next day. Reluctantly, the chief dispatcher gave his consent.

At precisely 4:00 p.m. a crew showed up, parked close to the corner, and for an hour and three-quarters sat watching the traffic. Absolutely nothing happened. Bored, they called up the dispatcher at a quarter to six and requested permission to clear. The dispatcher, fed up with having a car sitting on standby when there were a stack of transfers to be done, readily agreed.

As soon as the light turned green, the crew pulled away from the curb and headed into the intersection. At that moment, irony and Murphy's Law combined. A sports car tore through the red light to their left and—*SMASH!*—hit them broadside.

THE BREAKOUT

We had been sent to pick up a patient at the main police cells, and I had to nose the ambulance down a narrow ramp and around a curve to the basement entrance. On the way out again I had to back up the ramp and around the curve, and as I was unable to see much of where

I was going, three policemen waved me backwards onto the street. It was only as I turned onto it that I realized with a shock that I was facing into lane after lane of oncoming rush-hour traffic. I had forgotten that it was a one-way street, and here I was nose to nose with the traffic. The three cops were staring at me balefully.

I figured there was nothing to do but look legitimate, so I threw on the lights and drove straight into the oncoming cars with my sirens wailing. As drivers scrambled out of the way in confusion, I pulled ahead as fast as I could and got out of sight around the next corner. If I'd done that stunt in my own car, I'd still be paying off the fines!

RUNNING THE LIGHTS

It was late at night and we were clearing St. Paul's Hospital in downtown Vancouver when the dispatcher called. Someone in Burnaby was short of breath and we were the only available ambulance, so could we make it quick? My partner gunned it as far as the first red light, slammed on the brakes, accelerated through the intersection to the next red light, slammed on the brakes, accelerated to the next red light, slammed on the brakes, and so on all the way to Burnaby.

When we reached the house, the fire trucks were already there, lights flashing. We jumped out and ran into the house, only to find a party in progress and a young woman hyperventilating in drunken grief because her boyfriend had just dumped her. We gave her a brown paper bag to breathe into, told her to go home and get some sleep, and walked back outside.

The firemen were standing around our ambulance, dripping hoses in hand. "So much smoke was coming from under your vehicle, we thought it was on fire!" said one. "But it was just your brakes."

That was it. One emergency run, one set of brake linings. And all for nothing.

A Close Call

We had arrived back at the station a few minutes earlier, and I had just put the kettle on when the hotline rang. Josh picked up the line and jotted down the call number and address. Then he blurted, "Code 3 for a SIDS baby ... let's go!" While all deaths are sad, we are not surprised when an old person dies. But we don't expect children to die, and there is something particularly tragic about sudden infant death syndrome, the unforeseen and often unexplained death of a baby.

Josh and I ran down the stairs to the garage, slamming the door after us. In passing he hit the button to open the bay door and we jumped into the ambulance. He was driving and he reached down to turn on the key.

"Damn, where's the key?" He felt his pockets. "I must have left the keys in the station—but now we're locked out. Graeme, do you have a key?"

"No," I said. "I'm working spareboard. I don't have any station keys."

"Shit. We're going to have to break in to get the keys. One minute ... I think I left a window open on the second floor." He ran outside with me following closely behind. "Give me a boost."

I crouched down next to a drainpipe and Josh stood on my shoulders. Then I lifted him up and, with one hand on the pipe, he was able to climb high enough to grab the windowsill, yank himself up and dive headfirst through the open window.

Thirty seconds later we were back in the ambulance. Josh flipped on the lights and sirens and we raced onto the street. We were both tense, and I noticed that sweat was dripping from Josh's face onto his dirty shirt and the skin on his fingers was torn and bloody. He gripped the steering wheel tightly and drove like a maniac, barely slowing down as we ran red lights. We turned off the highway and tore down an exit ramp, the ambulance leaning sharply. But despite the squealing tires, we didn't roll, and soon we were skidding to a stop outside a small green house. I glanced at my watch: seven minutes had passed since the hotline had rung. Were we too late to revive the baby? Was it irreversibly brain dead?

We leaped out and I grabbed the jump kit and charged up the steps, Josh following with an oxygen bottle. The front door was slightly ajar, and as we ran inside, I yelled, "Ambulance! Where's the baby?"

A woman stepped out of a room down the hall. She was crying and holding a bundle in a white blanket; my heart almost stopped. Then she looked up at us and said, "It's okay. He's alive. The dispatcher told me to blow in his mouth and now he's breathing again."

I was so relieved! Even though Josh and I would not have been responsible for the boy's death, and although it was not my fault that the ambulance keys were left in the station, if that baby had died, I would be feeling guilty to this day.

Co-operating with Cops

In general, BC paramedics have excellent relationships with the other emergency services. Police and firefighters are often the first on the scene, and they deal with crowd control, provide critical first aid, extricate accident victims and help us carry patients and equipment.

Of course, even if 99.99 per cent of the time the calls go smoothly, that other 0.01 per cent of the time something will go wrong. And as you can expect, I'm going to tell you about those other times.

ONE-UPMANSHIP

The Vancouver City Police called us for a man down. We pulled up to the curb behind their paddy wagon and rushed over to a man who lay sprawled on the grass next to the sidewalk. Two cops were leaning against the wagon, looking unconcerned, their hands in their pockets. They were obviously enjoying the fine summer sunshine.

The man looked a little dirty, but otherwise healthy. He was young, tall and very tanned from living outdoors. I shook him. "Hi, wake up! Are you all right?"

The man stirred and opened his eyes. "Yeah, I'm all right. But who are yoush?" He reeked of wine.

"Do you hurt anywhere? Is anything wrong?"

"No—the only thing wrong ish that I'm thirshty," he said slowly. "Have yoush got anything to drink?"

"Sorry. We don't serve drinks in the ambulance. And it's too bad because we could make a fortune if they'd give us a liquor licence. Do you want us to take you to the hospital?"

"No. I'm tired. I jusht want to shleep. Goodbye." And with this the man lay back down on the grass and closed his eyes.

"Hey, wake up!" I prodded him. "You can't sleep here by the highway. If you roll over, you'll be run over. You have a choice. You can come with us to the hospital or go with the police and sleep it off in the drunk tank."

"Go away," the man grunted, never opening his eyes. "And good night."

I turned to the two cops. "There's no reason for him to go to the hospital, and he doesn't want to go. He's all yours."

One of the cops laughed. "No way. He's lying down. That means he's a medical case. You take him. We only take drunks who can stand up."

I was appalled. "You've got to be kidding! An emergency room bed costs thousands of dollars a day. It's stupid putting a drunk in a bed needed for a trauma victim."

"Hey, the whole system is stupid," the cop said. "These drunks should be locked up until they dry out instead of wasting everybody's time. But we didn't make the rules. If a man can't stand up, he's a medical problem." Both cops were grinning now, their hands still in their pockets.

"Okay, okay. We can play this game too." I was getting annoyed and the cops were bringing out my sense of male competition. "Hey, buddy—get up! You have to make a little trip." I shook the man awake again.

"Whazzat? Where am I going?" He made an effort to sit up. My partner and I grabbed his arms.

"We need you to help us. You're on our team, okay? You have to stand up. Yes, you can do it! Just push with your legs. That's it! Fantastic, you're up!" The man staggered to his feet, seeming pleased to be the centre of attention.

"There," I said, smiling at the police. "He's standing, and he's all yours."

But the cops didn't budge from the paddy wagon. One shook his head. "Not so fast. That guy doesn't look like he's standing to me."

The drunk had acquired the list of a sinking ship, and as we watched, he tilted right over, falling onto his side on the grass. Then he belched, curled up in a fetal position and went back to sleep.

"One minute!" I shook the man vigorously enough to make his teeth rattle. "You can do better than that. You've got to get up. You can't let your side down!" We tugged him back into a sitting position.

"Huh? You want me to get up? Oh, okay. Jusht for yoush." And with a little help he got back on his feet.

"All right, you guys. He's up again. Now take him away."

The closest cop was looking at his watch. "He won't last thirty seconds." And sure enough, the drunk started weaving around in wider and wider circles. Suddenly he pitched forward, landing flat on his face in the grass. In a minute he was snoring softly.

My partner looked at me. "We could try for the best out of five ..."

"Nah." I knew when I was beat. "The only way this guy will ever stay up is if we tie him to the stop sign. If he falls down any more, he'll get hurt, and that's not the kind of game I want to play."

We loaded him up and drove away, the cops waving cheerfully at us as we left. The nurses at emergency were predictably disgusted with having to waste a scarce bed on a drunk, and they left him to sober up on a gurney in the hallway.

HOT TVS

The cops had called us down to the main cells for a possible heart attack. A middle-aged man with a bit of a belly was sitting on a chair, groaning and clutching his chest. The sergeant called me to one side while my partner cracked open the oxygen tank. "I think he's just faking it to get out of here. He's a fence and he thinks we've got evidence on him, but actually we don't have a thing. You can take him away if you want."

We loaded him up and headed to St. Paul's Hospital. We weren't more than a block away from the cop shop when the patient perked up.

"That oxygen must be doing the trick!" he exclaimed. "I'm feeling just fine now." He pulled the mask off his face.

I put down my clipboard and asked, "So what kinds of stuff do you sell?"

The patient sat up straight on the stretcher. "I can get you and your buddy brand-new colour TVs tonight. And since you're being so good to me, I'll give you a special deal: two hundred bucks each for twenty-inch TVs with remote controls."

I laughed. "Two hundred bucks! What a rip-off. I can buy one on sale for that much." At this he became defensive. "No, you can't. But listen, if you aren't happy with that price, I'll knock it down to a hundred and fifty."

"A hundred and fifty!" I acted outraged. "You're getting them for almost nothing. Fifty bucks would be too much."

Now he was getting mad. "Do you know how much trouble we have to go through to get this stuff? It's not easy, you know. Maybe I could come down to a hundred and twenty-five, but that's it."

By this time we had arrived at St. Paul's, where the driver pulled up outside the emergency entrance. I looked at the patient grimly. "You've been wasting our time and you know it. There is nothing wrong with your heart. You're a fraud, you're a thief and you're an overpriced rip-off artist! We're handing you back to the police."

When our patient heard this, he jumped off the stretcher, opened the side door and ran off down the street. We yelled after him, "Police, stop him! He's going that way!" No police were in sight, but he ran faster anyway. We stood there laughing as he disappeared from sight. In this business it's a rare pleasure to see a cardiac patient recover so quickly.

DO *YOU* WANT TO GO TO HOSPITAL?

It was after eight o'clock on a dark, overcast night when we knocked on the Burnaby motel room door. A man in his thirties opened it and motioned us inside. Two women were standing by the TV, one of them with a badly blackened eye and a scared look on her face.

"What happened to you?" I asked.

"Oh, nothing," she replied.

Before I could ask her another question, the man interrupted. "We didn't call you to see her. I'm worried about the guy down there."

He pointed, and I turned to see a large body stretched out between the two double beds. The man on the floor was also in his thirties, but taller and powerfully built.

"So what's wrong with him?" I moved closer to the prone figure and put my jump kit down on the nearest bed.

"He drank a forty-ouncer of vodka and passed out. That's not normal. He should be able to drink more than that."

I leaned over the body and shook his shoulders. "Hey, wake up! Are you all right?"

The man opened his eyes and looked at me warily. "Are *you* all right?" he responded.

I moved back a half metre. "Do you want to go to the hospital?"

The man twisted around and staggered to his feet. "Do *you* want to go to hospital?" He clenched his huge fists and started toward me.

"Hold on there!" I grabbed my kit and backed up slowly. "We're just here to help you. If you don't want help, we'll be happy to go on our way." But now the man was between me and the door, and beyond his shoulder I could see my partner slip outside. So much for the buddy system, I thought. Now what? I didn't stand a chance against this monster.

Then the man's friend said something, and the brute turned to talk to him. I grabbed the opportunity to dart past him, out the door and into the ambulance. My partner, already in the driver's seat, quickly backed up twenty metres.

I was angry. "Where the hell did you go? That guy could have killed me!"

"Yeah? He would have killed me too so I called the police. They should be here any minute."

Sure enough, a squad car turned up a minute later, but with only one cop in it. He pulled up beside us and climbed out to look at the motel room. By now the man was standing in the doorway, shouting obscenities. "You pig, why don't you take off your gun and fight like a man? You coward, come closer! I'm gonna rip your arms off. Just come over here!"

The cop didn't move. "Alex Campbell," he said. "I thought it would be him. He was released from the pen two days ago. He's spent his whole life in and out of jail for assault, armed robbery, rape, the whole works."

"Well, aren't you going to arrest him now?" I demanded. "He's drunk and dangerous. I'm sure he's beaten up one of the women in there."

"No, I'm not going near him," the cop replied coolly. "I'm not getting my neck broken just to have him released by a magistrate tomorrow." And he leaned down to give a report over his car radio.

I couldn't believe what I was hearing. "But what about that woman? You can't just leave her there at his mercy."

"She could have left while he was sleeping. She'd probably go back to him even if we got her out of there. I'm not going to try to sort that one out right now. No, we'll just wait until he pulls a weapon on us. It won't be long. Then we'll gun him down and be rid of him."

We Code X'd the call and drove away. It's a rough world out there, and when it gets really dirty, the rules don't matter a damn.

A REAL BIND

Working spareboard in the Lower Mainland meant not having a regular station or partner. On almost every block, I was sent to a different station to fill in for a crew member who was sick or on holidays. During this time I was paired with many paramedics, and I got to know many of them quite well. Invariably, at some point during our long shifts, our conversation would turn to our job, and my partners would tell me about some of the exceptionally challenging or strange calls they had attended. Like this one...

An obese man had died several hours earlier in his second-floor flat, and now the heavy body lay face up on the bed, legs spread wide apart and stiff with rigor mortis. The ambulance crew looked at each other in dismay.

"How do we get him out of here?" the driver asked his partner.

"The easiest way would be to push him out the window, but we'd probably squash someone down below." A small crowd had gathered on the street to see what the ambulance and police cars were up to.

"Well, let's ask the two cops in the hallway to help. If we strap him onto the scoop, the four of us can carry him down the stairs, but we'd better cover him up really well with blankets to avoid offending anybody. You never can tell when the press will show up with cameras."

They broke the scoop stretcher apart and snapped it back together under the massive corpse. Then they lifted up the stretcher, with a male paramedic and a female cop at the front, and another paramedic and cop at the back, and lugged it to the head of the stairs.

The lead paramedic looked dubiously at the stairs, which made a ninety-degree turn halfway down. "We'll never make it around that bend carrying him this way. His legs are sticking out too far. We'll have to lift him up to our shoulders." Grunting, the four heaved the blanket-covered mountain onto their shoulders and started down the steps. Below them, faces stared curiously up at them through the open outside door.

Halfway down, the straps started to shift over the corpse's rolls of fat, and he began to slide forward on the tilted stretcher. "Hang on!" the cop at the back shouted, and they all stopped, hands gripping the scoop's metal frame.

But the body, legs forming a stiff V, continued to slide until the cold flesh forced the heads of the paramedic and the female cop together at the front end of the stretcher. Finally it stopped, resting against their necks. Their cheeks touching, the two grimaced and looked out of the corner of their eyes at each other.

"You know," the policewoman gasped, "we're going to have to stop meeting like this!"

At this they stood still, shaking with a combination of hysterical laughter and sagging knees, while the paramedic and the cop at the back of the stretcher tried desperately to keep the whole show from collapsing down the stairs and into the waiting crowd.

Working with Firefighters

Firefighters often back up ambulance crews on emergency medical calls.

We were sent Code 3 to an address in East Vancouver for an unknown problem. By the time we arrived a fire truck was already parked outside the house. The front door was open and we followed the sound of voices into the kitchen, where a fireman was trying to talk to a deaf old man. The fireman almost shouted, "Do you have any medical problems?"

The old man looked annoyed. "Of course I have medical problems!"

The fireman said, "Is anything out of the ordinary bothering you?"

"Out of the ordinary? Do you think I called you because I've got crabs?"

I have a lot of respect for firefighters. I was once part of an experiment where we observed firemen while they were battling a fire burning in a six-storey office building on a hot summer day. They were wearing heavy turnout gear while pulling charged hoses up ladders and stairs, and they were searching for possible victims in hallways and offices where the smoke was so thick they had to feel their way and where at any time the floors or the walls could cave in and kill them. Because they were sweating litres of fluid, every twenty minutes or so they had to come down to rehydrate and take a brief rest. That was when we checked their vitals and found that many of them had pulses over 150 beats per minute and systolic blood pressures around 200. That job is no picnic!

THE MYSTERY OF THE CRUSHED BUG

Before BC's Ambulance Service was created, some of the larger fire departments had employed paramedics, and many fire chiefs and union officials were not happy to lose funding and staff to the new service. This resulted in some tense incidents in the early days of the service.

The call was Code 3 for an accident on a Vancouver bridge. "Code 5s and 6s will also be attending."

Brett deftly wove his screaming ambulance through the early afternoon traffic, but four minutes later when they arrived on the bridge, they found the way blocked by a long red pumper truck parked sideways right across the lanes. A broad-shouldered fire chief held up his hand. "Stop!" he yelled.

Brett stepped out of the ambulance, his partner Warren following with the jump kit. "What's going on?" Brett asked.

"There's been a bad accident," the fire chief replied. "We're checking it out. We'll let you know if you're needed."

"No, no," Brett protested. "It's our job to see whether anyone's hurt." He started to walk past the fireman but was blocked by a raised arm.

"Not so fast," the chief said. "You're not going in unless I tell you to."

Brett wasn't as wide as the older man, but he was taller and he wasn't backing down. He pushed the arm away. "Yes, we are!"

But the fire chief shoved him back. "Stay where you are. I'm in charge here!"

Up to now Brett had been annoyed; now he was furious. He balled his fists and shouted, "Get out of my way!"

Just then a policeman showed up. "What the hell are you two doing? What's the problem?"

The chief said, "I'm in charge of this scene, and I'm telling this young punk to stay where he is and not interfere with my men."

The cop glanced up at the accident. "You're in charge? I don't see any fire. This is an accident scene, so you're not in charge, I am. Now settle down before I arrest both of you for causing a disturbance." By now cars were backed up around the corner and people were getting out to see what was going on. "Everyone move back!" the cop shouted. "Don't crowd the emergency vehicles!"

While Brett and the fire chief were arguing, Warren had slipped around the far end of the pumper truck, where he joined four firemen who were peering at the tangled wreck of a bright yellow Volkswagen Beetle. The driver must have been going much too fast and lost control because the car had crossed over into the oncoming lane and hit a large truck. Its front end was caved in, and its frame was so badly buckled that the driver's seat was pressed against the steering wheel and the rear seat and engine were squeezed against the front seats.

One of the firemen said, "This is really weird. This Bug must have had a driver, but we can't see a body in there!" As the windshield was missing, Warren leaned through the gap. "Could someone be trapped under the dash?" The fireman shook his head. "That's not possible. We'd be able to see a foot or some blood or something. We'll cut it apart. Maybe we'll find a body under the back seat, but I doubt it."

Warren stood and looked around. A thin man with unkempt hair was standing on the sidewalk, staring at the wreck. Warren approached him and asked, "Did you see the accident?"

The man nodded.

"What happened?" Warren asked.

The man nodded again and then again.

That was when Warren noticed that his face was bruised and he was shaking. Suddenly he understood. "Is that your car?"

The man kept nodding.

"Brett!" Warren called. "Here's our patient—let's get a neck brace on him and put him on a backboard."

It's almost always safer to wear a seatbelt because people who don't are thirty times more likely to be ejected from a vehicle in an accident, 50 per cent more likely to suffer a serious injury and 45 per cent more likely to die. But there is always a freak exception to every rule. In this case, at the moment of impact the driver had been ejected through the windshield at just the right angle to avoid smacking into the oncoming truck and with just enough force to land him on the sidewalk rather than striking the bridge railing or tumbling right over it into the ocean.

That was one lucky guy! He must have had nine lives. At least he did before that accident.

DEEP FRIED

One of our heart patients had a cardiac arrest while we were lifting him into the back of our ambulance, so two firemen climbed in to help us with CPR (cardio-pulmonary resuscitation). One held a rubber oxygen mask to the patient's face and squeezed an Ambu bag to ventilate his lungs with oxygen while the other pressed down on his heart to circulate his blood. After a minute we asked the fireman at the head to stop while my ALS partner intubated the patient. Then I attached the bag directly to the intubation tube and the fireman returned to ventilating.

Next, I prepared to defibrillate the heart: I put gel on the two paddles, rubbed them together to spread it around and waited for the monitor to indicate that it was fully charged. I asked the second fireman, Sam, to stop compressions and placed the paddles on the patient's chest. "Clear!" I ordered, and after a few seconds pressed the discharge button.

The patient's body jerked, and Sam yelled, *"Yow!"* and began rubbing his legs. "Ow, ow, ow!" he moaned. "I got zapped. I guess my knee was touching the stretcher."

The other fireman laughed. "I guess you can forget about making love for a month or two, Sam."

I turned back to the monitor: still no viable heartbeat.

My partner said, "Let's give him another minute of compressions, and if there's no improvement, give him another shock."

There was no change, so again I called out, "Clear!" but this time before I pressed the discharge button, I took a brief look at Sam. No worry—he was holding his legs as far from the metal stretcher as the narrow aisle allowed. Some lessons you never forget!

THE JAWS OF LIFE

Somewhere I have a yellow certificate stating that I attended a course in vehicle extrication. The course teaches paramedics how to free people trapped in smashed vehicles, and for a couple of days we learned how to break windows, pry off doors and peel back roofs without causing further injury to accident victims. Although these skills came in handy on a number of occasions, whenever possible I left the rescue work to the experts—the firefighters. After all, they had more equipment and experience than we did, and chopping and dicing metal is hard work.

On one such occasion an elderly man was driving along the Trans-Canada Highway when he had a stroke. Unconscious, he let go of the steering wheel and his car veered onto the shoulder and into the ditch, where it came to a stop and stalled.

We could see the accident scene from a distance. Two red fire trucks were parked at the side of the road, and a small crowd had gathered to watch the firemen unload a large assortment of rescue tools from their vehicles, including hydraulic jacks, wooden cribbing, cutters and spreaders. Two of the firemen were clearly preparing to pry off the driver's door with the Jaws of Life.

We parked behind the fire trucks, picked up our equipment and hurried to the busy scene. A boy about ten years old who was standing on the far side of the car waved furiously to us as we approached. My partner and I climbed down into the ditch and went to see what he was pointing at.

Although three of the car's four doors were locked, the boy had discovered that the front passenger's door was unlocked. Fortunately, the ditch was wide and the car was sitting upright, and my partner and I were able to force the door open far enough to get inside, examine and treat the patient and extricate him on a backboard. We carried him out of the ditch and loaded him into the ambulance.

Just before I closed the back door, I looked back at the accident scene. To my amazement the firemen were still clustered around the car, busy cutting off doors, seemingly oblivious to the fact that the driver was gone. I was baffled. How could six firemen not have seen us crawl inside the car and remove the patient? I was too busy stabilizing the patient to think about this until after we'd handed him over to the emergency room staff. Then the answer came to me. Obviously the firemen had realized their mistake after we opened the door on the passenger's side, but having started, they weren't going to miss the chance to practise with their fancy rescue equipment.

I would have loved to see the expression on the insurance appraiser's face when he went to the wrecker's and saw that car. I'm sure nothing was left but a bare chassis and a pile of metal scraps!

A Helping Hand

One of the Vancouver ALS crews used to carry around an extra drug box filled with sandbags. At calls, Vince would ask a young fireman to go to his ambulance and fetch that second drug box—the big red one in the back cupboard. As soon as the fireman had lugged it up several flights of stairs, Vince would tell him to take it back to the car as it was no longer needed.

I found this joke offensive, so back at the station I went up to the crew and protested. "It's pretty mean to play a trick like this on the firemen when they're doing their best to be helpful."

Vince grinned. "Who's being mean? Those young guys want to look good for the girls. I'm just helping them stay in shape!"

THE MAN WHO CRIED WOLF

If you are homeless in Canada, perhaps because of addictions or other mental health or financial problems, you might want to move to the West Coast. Although living on the streets in Vancouver is no picnic—it can be cold, wet, dirty and dangerous—it rarely snows and the emergency services are first class.

Alphonse knew all this. A sixty-something alcoholic with a red face and puffy nose, he had lived in Vancouver for years, sometimes sleeping in shelters but more often in doorways or under park bushes. During the day he scrounged for spare change then spent most of it on cheap wine. He would drink until he passed out and often awoke in a police cell or hospital bed.

After a few years he was so familiar with all the emergency services that he was an expert on the best way to play the system. His goal was to end up in an emergency room bed where kind nurses would provide a thorough checkup, a warm meal and a change of (used) clothes. If he was very lucky he might even be allowed to sleep off his hangover on a spare cot. His main challenge was to avoid getting picked up by unsympathetic cops and tossed into the drunk tank instead.

So Alphonse worked out a routine. Twice a week he would show up at a fire hall, ring the bell and sit down on the steps. When a fireman opened the door, he would complain of chest pains. Following standard procedure, the firemen would then call an ambulance. Although the monitor would indicate that Alphonse's heart was fine, the paramedics could not ignore his complaints and would have to transport him to emergency. There, the triage nurse would have to classify him as a possible cardiac patient and immediately transfer him into a soft hospital bed.

After a month of this, everyone—except Alphonse—was completely fed up. The firemen told him to get lost, but after years on the streets he was immune to verbal abuse. Nobody was going to keep him away from a nice warm bed. The next weekend he went on his usual three-day binge. Afterward he was in rough shape—cold, wet and hungry—so by Monday afternoon he was back at the fire hall. When he rang, a fireman opened the door.

"I've got chest pains," he said. "I need an ambulance!"

"One minute," the fireman said. He went inside and told the fire chief that the old drunk was back. The chief was furious. "Tell him we've had it with his games. He can freeze in hell before anyone here is calling an ambulance."

The fireman went back. "You might as well leave. No one here is going to give you any help."

"But you have to," said Alphonse. "That's your job!"

"No, it isn't. My job is to fight fires, not coddle winos."

"Well, I'm not leaving. I'm staying right here until you call an ambulance." And Alphonse lay down in front of the door. Soon he was fast asleep.

The fireman told his boss.

"Fine," said the chief. "Let him lie there. Tell everyone to completely ignore him. He'll go away eventually."

So Alphonse stayed. At 6:00 p.m. the shifts changed, and the night crews stepped over him on their way into the station. Late in the evening someone felt sorry for the old alcoholic and covered him from head to foot with a blanket. When the day shift came back at six the following morning, Alphonse was still there and the crews stepped over him again. "Is he still sleeping? He must have been on a real bender!"

But by late afternoon one of the firemen was getting worried. He pulled the blanket back. The old man's face was yellow and cold. Sometime over the last twenty-four hours he had suffered a massive heart attack and died.

I don't know how the fire chief explained that one. The paperwork must have been awful. I really pitied him, though perhaps I should have had more sympathy for Alphonse. Who knows why he became an alcoholic?—maybe he was neglected as a child. Obviously his parents had never read him the story of the boy who cried wolf.

What Makes a Paramedic Throw Up?

The trick to keeping a cool head and a calm stomach in this paramedic business is to keep a professional distance from the patients' pain. You can do your job as long as you remember that you are treating the injury, not suffering from it. Lose that distance and you are in trouble. You also need to learn how to put difficult memories in a mental closet and close the door. But even knowing all this, over the years some situations did get to me.

The first time it happened to me was back in Edmonton when I was an operating room attendant. I had been asked to hold an elderly man while a surgeon did a spinal tap. The doctor inserted the huge needle slowly into the patient's lumbar spine. Then he withdrew it. He had missed. He swore under his breath and reinserted it. He missed again. He tried again. And again. Each time the patient groaned with pain.

Suddenly I felt nauseated. My knees started to buckle. "Sorry," I said, gasping. "I'm going to faint."

A nurse shoved a stool under me and caught me as I fell.

At that point the doctor lost it. "Who is this incompetent idiot?" he shouted. "Get him out of here. I never want to see him again!"

The nurse helped me to step outside the OR. As she turned to go back in, she grimaced and said, "That doctor is terrible!" Fortunately, he wasn't typical.

When I first started working as an emergency medical attendant, I held the older paramedics in awe. Some of them had three or four bars on their jackets, signifying fifteen or twenty years of service. They must have seen everything! So I decided to ask them whether they had ever had a call that made them sick. I'll never forget some of their stories.

THE INJURED BRAKEMAN

The train had jolted backwards without warning, pinning the brakeman between two freight cars. The ambulance crew expected the worst and were surprised to find the man standing up, leaning against a boxcar and having a smoke. Dwayne asked him how he was doing.

"Not much pain. But I think I've hurt my knee." There was a small patch of blood on his pants below his left knee.

Dwayne said, "If you don't mind, I'll slice open your pant leg so I can have a good look."

"No problem, go ahead."

The paramedic cut away the fabric, expecting to see a minor wound. Underneath, most of the knee joint was missing. Dwayne was so surprised he threw up.

OVER THE TOP

Brad wasn't feeling well. He had partied hard the night before, awoken late with a splitting headache and missed breakfast. Now the shift was starting with a Code 3 for a vehicle that had gone off a mountain road and rolled down a cliff.

It was a difficult descent to the wrecked car. As they drew close, they could see that its frame was badly bent: the back seat had been folded forward and the front seats pushed toward the dash. The driver was dead, impaled on his steering wheel. Beside him was the body of a woman. She was headless, decapitated by the windshield.

To this day Brad cannot completely explain why he'd vomited. "I've seen lots of trauma, but this was the only time it really got to me. Maybe because I was hungover."

He was standing next to the car, retching with loud, painful heaves, when suddenly he heard a small squeak. His nausea disappeared and he yelled to his partner, "Tom, call for the Jaws! Someone's still alive in there!"

The firemen cut the car apart and found a six-year-old girl lying hidden on the floor underneath the folded back seat. Although she was unconscious and had broken ribs and fractured legs, she eventually made a full recovery.

If Brad hadn't thrown up, the girl would not have been roused from her coma and made a noise. In all likelihood no one would have known she was trapped until it was too late.

They say God works in mysterious ways.

The Laws of Probability

It's strange how paramedics can go for months without seeing anything dramatic, like a bad car accident or a cardiac arrest or a murder, and then there's a run of them in a short period—even two or three back to back. It may just be the laws of probability, but sometimes it seems that the stars align—or misalign—and things start going crazy.

Dawn and Bob looked like a mismatched crew. She was short and stocky; he was tall and lanky. She was married with two kids; he was single and a party animal. Nevertheless, they had been partners for years because they worked well together, and both of them loved their jobs. One day they were called out for a suicide—a jumper—in downtown Vancouver. The body was lying face down on the sidewalk, and although Bob was pretty sure the man was dead because a bystander said he had fallen four storeys, he bent down to check for any sign of life. Just then the guy's brother jumped from the same balcony and landed two metres on the other side of him. It was such a weird and unusual event that in a couple of hours everyone had heard about it. Of course, some of the guys started making jokes, and by the next day a fake Workers' Compensation Board safety notice had been posted in the station to warn emergency personnel not to stand close to Bob anywhere in the vicinity of tall buildings.

The rest of Bob and Dawn's week passed uneventfully, but on the second day of their next block of shifts they got a Code 3 for an attempted suicide. They walked into a living room to see a thirteen-year-old kid sitting in a large pool of blood. In front of him was his left hand, which he'd hacked off with a meat cleaver. He was still alive but in shock, so they whistled him into emergency, but the surgeons couldn't reattach his hand because his wrist was too badly damaged.

Bob said he'd never felt sorrier for anyone in his whole life. By the end of that shift he and Dawn were feeling uncharacteristically sad and subdued. The other crews noticed, and Pete reminded them, "Don't forget, this is our platoon's pub night. The two of you look like you need some fun."

Dawn caught Bob's eye and he nodded in agreement. "You're right," she said, "we could use a few drinks. I'll talk my husband into going. It's too late to find a babysitter, but we can probably drop the kids off with Granny and Grumpy. See you soon!"

The bar was crowded and noisy, with a Canucks game playing on the big wall TV. Fortunately one of the paramedics had come early and secured a long table, and by 8:00 p.m. twelve people were downing pitchers of beer and shouting to be heard over the din.

Pete said, "Talking about traumatic situations, did you know that a woman with a PhD in psychology has been hired to find out how emergency personnel cope with post-traumatic stress? She asked J-C whether he had a technique for coping with stressful incidents and he said, 'Yeah, as soon as I get home I jerk off till I pass out.'"

Everyone except Jeanette laughed. "What a stupid, offensive thing to say," she said, "especially to a woman! Do you want her publishing a report that says paramedics manage stress by masturbating?"

"Oh, come on, Jeanette, chill out," Leo said. "J-C was just joking. And I'm sure there's not much anyone could say that would upset a psychologist."

"That's the problem with J-C," Jeanette replied. "To him everything is a joke."

"What do you want him to do?" Nate yelled from the end of the table. "In this job if you can't learn to laugh, you'll be crying all day long."

Dawn interrupted. "That reminds me of a call we had last year. Bob and I were backing up Leo and Hans on a possible Code 4, and when we got to the apartment, we were met by this really big middle-aged woman. She was frantic and kept saying, 'Hurry up, my father's not breathing! You've got to do something! Hurry up and do something!' Her father was lying on his side with his face away from us, so Hans turned him on his back to get a better look, but when he did, the man's arm shot up in the air and stayed there like he was waving goodbye! The old guy was in rigor—he must have been dead for hours. Leo muttered to Hans, 'If you can restart this one, I'll call you God,' and Bob lost control and started giggling. Well, that really set off the daughter. She was so mad she almost punched him out. She could have too—she was built like a prize fighter. It was so funny! Remember, Bob?"

"Uh, yeah," Bob said distractedly. "Excuse me, I gotta go." He got up and started weaving past the noisy drinkers and the waitresses with loaded trays.

Pete was worried. "Is Bob okay?"

"I think so," Dawn said.

Bob was heading toward an attractive woman sitting by herself in the far corner. By the time he returned, the others had ordered more pitchers of beer and were talking about the hockey game. "I'm just getting my jacket," he said. "I'll see you tomorrow at five." He waved goodbye. Behind him the attractive woman also gave a friendly wave.

When the three crews in Bob's platoon showed up for the night shift the next afternoon, they were all curious to learn about the mystery woman. Jeanette asked, "What's her name?"

"Meredith," Bob answered. "She's a lovely woman."

"Have you known her long?"

"Nah, we just met last night. I smiled at her, she smiled at me, and we got together."

"Where did you go?"

"If you've got to know," Bob said, "we went to her place and spent a wonderful night together."

Dawn grinned. "That sounds pretty romantic! Is this the beginning of something serious?"

Bob shook his head. "Not likely. Her husband is coming home tomorrow."

"What?" Pete said. "She has a husband? What else does she have, herpes?"

Bob smiled. "I don't think so. At least she said she didn't."

"And you trust her—a woman who's fooling around on her husband?" Pete was getting riled up. "Don't you know adultery is a sin?"

"Don't worry, we were just having fun. And if I'm going to sin, I prefer sins of commission to sins of omission. It was a fantastic night. I would've never forgiven myself if I had turned it down for moral reasons."

Pete said, "When it comes to sin, you don't get to forgive yourself. That's up to a Higher Power."

Now Bob was starting to get pissed off. "When did you discover religion, Pete? You haven't stepped inside a church in twenty years."

Leo laughed. "Pete's not religious, he's jealous. He hasn't gotten lucky in six months. But tell us, what's your trick? Why is a thin, bony guy like you successful with women?"

Bob chuckled. "There's no trick. Just be nice and friendly. And of course, if you don't ask, you don't get."

Dawn said, "Bob's trick is good bedside manners."

Leo was confused. "Bedside manners?"

"Bob is always polite, kind and considerate. For example, on our last recertification when we had to demonstrate our IV skills, when Bob stuck a big needle in the rubber arm, he said, 'There, there ... now that didn't hurt, did it?'"

"How sweet." Pete scowled. "I bet he said the same thing to Meredith last night."

"See?" Leo said. "I told you Pete's jealous!"

She Can't Die Here!

We were cruising down Broadway, feeling pretty mellow on a crisp, sunny Christmas Day. But dispatch interrupted the calm: Code 3 for a suicide attempt in Kitsilano.

At the apartment block, we took the elevator upstairs, where we were met in the hallway by an older woman dressed in her Christmas best.

"What am I going to do? My neighbour wants to kill herself in my rocking chair!"

"Don't worry, ma'am. We won't let her do that."

In the living room a younger but more haggard woman was clutching the arms of the chair, her eyes on the floor. "It's not worth it any more. I just want to take some pills and die."

"Oh, come on, dear," I said, using my most reassuring tone of voice. "When you're feeling lonely and down, everything looks hopeless. Tomorrow when you start feeling better everything will seem wonderful again, and you won't even be able to remember how sad you felt today."

"Don't give me that garbage! There's no point trying to cheer me up. It never gets any better for me. I'm just going to stay here and kill myself."

At this the older woman threw up her hands. "I've got all my Christmas guests coming for dinner in fifteen minutes. She can't just stay here and die!"

"Relax," I said. "She won't die that quickly." I turned back to the haggard woman. "Now, dear, just come along with us to the hospital and let the doctors give you some help with your depression."

"No. I don't want any help and I'm not going in your ambulance. Just leave me alone!" She started rocking back and forth in a short, agitated rhythm.

"Well, we're not allowed to kidnap you if you don't want to come with us, but you can't stay here and ruin someone else's Christmas. If you want to die, you have to do it in your own apartment. Now come on, get on our stretcher and we'll help you get back down the hall."

"All right," she said. "You can take me back to my apartment. But I'm not going to the hospital!"

We wheeled her back to her apartment and took her inside. There I made another appeal. "At the hospital you'll be taken care of and have a nice Christmas meal. You don't have to stay here alone. Come on, let us take you there."

She was weakening. "Okay, but you'll have to bring my slippers, my toothbrush, my glasses, my housecoat and my purse."

"Don't worry," I replied. "My partner will round up all that stuff while I make a list of your medications." I opened the door to her bedroom and stopped in surprise. On every surface were bottles of pills—pills of every description and size. Hundreds of bottles. She must have gone to a different doctor ever week complaining of different problems, been prescribed different medications and not taken any of them.

We couldn't leave all these drugs behind. They had to be sorted out and the extras thrown away. I found two big plastic bags and filled them. Then I got an extra pillowcase from under the head of the stretcher and filled it. Finally, my arms full, I walked back into the hallway. "I've run out of bags, and she's got even more pills." It was so absurd I couldn't help grinning.

The woman saw my smirk. "You frauds! You don't care about me at all! You think it's funny if I die. Well, I'm not going to any hospital. I'll just stay here, take some pills and kill myself." With surprising agility she hopped off the stretcher.

"Fine," I said. "Suit yourself. Here are all your pills back. Now have a Merry Christmas!" And I herded my bewildered partner out the door.

"Are you sure we can just leave her there?" he asked. "What if she overdoses? Shouldn't we call the cops and have her hauled off to a psych ward?"

"Are you kidding?" I said, laughing. "She hasn't taken a pill in ten years. She loves being sick—it's her whole life. She's not going to kill herself, and I'm not going to ruin her Christmas by trying to make her feel better."

And I was right. She didn't kill herself and we were never called back.

Hurtin' Country Music

Herb had loved country music all his life, and he promised himself that as soon as he retired, he would go to Tennessee and hear some of the greatest singers performing at the Grand Ole Opry, the home of country music. So he booked tickets well in advance, and the week after he turned sixty-five he flew to Nashville and checked into a hotel a few blocks from the concert hall.

His first show started at 7:00 p.m., so at six Herb put on his best cowboy boots and went downstairs to the bar. There he warmed up with a few drinks before walking out to the street. By 6:30 he was standing across the road from the Grand Ole Opry, full of excited anticipation. Distractedly thinking about the show, he stepped out onto the street and *wham!*—got hit by a big Ford pickup. Twenty minutes later he was in an emergency room being treated for a compound fracture of his left femur.

The next day he discovered that his BC Medicare wasn't valid in the US. The bill for his leg surgery was $17,000, and every day he stayed in hospital would cost him another $2,500. The cheapest solution was to fly back to British Columbia, but to do this, he had to buy tickets for three rows of seats because the airline would have to remove the middle row to accommodate a stretcher. He also had to hire a flight nurse to take care of him on the trip home.

We were sent to pick him up at the Vancouver airport. We parked the ambulance on the tarmac next to the plane, and after the rest of the passengers had disembarked, a ground crew rolled a gantry up to its rear door. We went inside, lifted his portable stretcher off the back seats and carried him down the steps. We were just about to place his stretcher on top of our main cot, when Herb yelled, "Stop! First put me down so I can kiss the ground. I'm so happy to be back in Canada!"

We set his stretcher on the ground, but because we couldn't flip him over, he kissed his hand then touched the ground. "I'm not going down there again," he told me as we drove to the hospital. "From now on, if I want to hear a top American country singer, I'll wait until they visit Vancouver."

I guess no one had told Herb that you should buy private medical insurance before you visit the US. For all its problems, Canada's health-care system provides everyone with affordable care, so it's easy for Canadians to forget that in some countries it only takes one illness or accident to send you to the poorhouse.

Miss-understanding

We were called to a recreation centre for a woman who had fainted beside the pool. When she saw us pulling the stretcher through the lobby, a five-year-old girl ran up to us.

She asked, "Is somebody hurt?"

Without stopping, I replied, "Someone's sick by the pool."

Trotting beside us, the girl said, "Someone sinked in the pool?"

I repeated, "No, we think someone is sick—sick, sick!"

The little girl looked confused. "Someone thick sinked, you think?"

Down and Desolate

Paramedics are frequently exposed to the dark underbelly of society—the many homeless, hungry and frightened people living on the margins because of addictions and other mental health issues, or simply because of bad choices, hard luck or poverty. Most of us try to avoid talking with homeless people when we pass them on the street, especially if they are dirty, pungent or obviously disturbed or drunk. In my experience there is usually little to fear. Most of the alcoholics and addicts I attended were friendly enough, even if they were badly confused, like the drunk who climbed into a cop car, thinking it was a taxi. The police dog inside bit him on the arm, and we had to take the man to emergency for repairs.

Of course, communicating with confused or inebriated patients can be challenging. One of my partners tried a new pitch: "Trust me, I'm from the government and I'm here to help you." But the drunk looked at him suspiciously and said, "No, no. I just want to go to a hospital."

Not everyone sleeping on the streets has mental health problems. Many people just can't afford housing, given that it's impossible for anyone earning the minimum wage to rent an apartment in large Canadian cities. And many people who suffer from mental illness don't live on the streets: a lot of depression and desperation is hidden behind the pleasant façades of comfortable high-rises and homes.

THE LONELY HEARTS CLUB

On Friday and Saturday nights, downtown Vancouver would go crazy with crowds of exuberant, laughing, screaming people. The bars would be full, and we rushed around picking up tipsy men and women who had fallen off stools, had broken bar glasses shoved in their faces or had driven into the nearest lamppost. Fortunately, Vancouver is in Canada, and though lots of fights occurred, there were few fatalities because not many people carried guns.

One of our Saturday night calls came in the wee hours as a possible MO [mental observation]. It was for an apartment in an expensive West End building, and we found a tall, thin man of around thirty sitting at his kitchen table. On it were a couple of empty whisky bottles and what looked (and smelled) like the remains of half a dozen joints.

"Hi," I said. "What's the problem?"

"I'm suffering," he moaned. "My lover left me, the cheap bitch! Now I'm all alone." At another time he might have been handsome, but now his thick brown hair stuck out in every direction, his shirt was stained and tears ran down his cheeks.

"I'm sorry to hear that. Why did you call us?"

"Because I'm lonely! I need someone to talk to. You look like nice guys. Sit down and have a drink." He reached for a bottle but missed.

"I'm afraid you're out of luck. We're emergency paramedics, which means we have to be available for emergencies like heart attacks."

"But this is an emergency!" he said, starting to sob anew. "My heart is broken in little pieces." Suddenly he fell to his knees and grabbed me around the legs with surprising strength. "Don't go! I need you!"

My partner laughed but I wasn't amused. This lunatic was slobbering all over my freshly dry-cleaned uniform pants.

"Look," I said, barely containing my annoyance, "you've got two choices. We can leave you here or we can check you into a mental hospital where you can talk to a psychiatrist. Take your pick. Stay home or get locked up."

The man let go of my legs, and for a moment he seemed almost sober. "No. That's all right. You can go." He staggered to his feet and tried to look dignified. "Thanks for coming. Have a pleasant night."

My partner opened the front door and pulled the stretcher out into the hallway. I took another look at the drunk, wondering whether it was safe to leave him alone in the apartment. How unstable was he? Any chance he might try to kill himself?

Then I said, "Get some sleep. You'll feel better in the morning," and I closed the door.

It's a hard world out there. Sometimes I wonder whether I made it harder.

THE MAN WHO COULDN'T DIE

Nate bought half a dozen bottles of over-the-counter painkillers, swallowed them all and lay down to die. He woke up in a hospital emergency room with a splitting headache, a sore throat and nausea.

A male nurse said, "You should be okay—we've pumped out your stomach. You're lucky your daughter was worried about you and called the ambulance."

"Who's lucky?" Nate groaned. "I'd be better off dead and my family would be better off without me."

He was kept in the hospital for three days and seen by a psychiatrist who prescribed antidepressants, arranged for him to get counselling and sent him home.

For Nate, the failed suicide attempt was another humiliating experience in a string of humiliations. He had worked as a bookkeeper for the same company for twenty years but was passed over for promotion in favour of a younger employee. After that he didn't work as hard, and when the company cut staff, he was made redundant. He then discovered how difficult it is to find a good job when you are fifty-five years old, and for several years he bounced from one part-time job to the next. In the process he lost his self-esteem and his confidence, became depressed and started spending his evenings in bars. One night he staggered home late to find a note from his wife saying that she had left him in disgust. He couldn't blame her. He disgusted himself. That was why two months later he decided to end his life.

Nate had narrowly escaped death, but instead of seeing this as an opportunity to make a fresh start, he saw it as proof that he was a loser. Why bother with counselling—it wouldn't give him back his job, marriage or self-respect. He was now even more convinced that dying was the only way out, but this time he would make sure that no one could bring him back to life. He would poison himself, hang himself and shoot himself all at the same time.

So he bought more pills, tied a noose to the chandelier in his apartment, pushed a chair underneath it and loaded his rifle. He swallowed the pills, climbed on the chair, put the noose around his neck and pointed the gun at his head. Finally, he kicked the chair out from under his feet, simultaneously pulling the trigger.

Two hours later he awoke in the same emergency room feeling nauseated. The only difference was that this time he had a bandage on his head, his headache was worse and his throat hurt both inside and out. The same nurse leaned over him and said, "You have to stop doing this!"

Nate was confused. "Why am I here again?" he asked. "What happened?"

It seemed that when Nate kicked the chair away, the movement threw his aim off so that the bullet just glanced off his skull. Then the weight of his body ripped the chandelier off the ceiling, and it—and Nate—fell to the floor. When the neighbours in the apartment below heard the shot and the crash, they called 911, and within five minutes the paramedics had removed the noose from his neck and were giving him oxygen. Fifteen minutes later the ER staff were flushing the drugs out of his stomach.

I hope he got the message that it wasn't his time to die. His daughter obviously loved him, so there was hope for a happy ending. And when you're really down, the only way you can go is up.

Another Rainy Day

I was sent to work on 13 Bravo on a typical overcast winter day in Vancouver, cool with a steady drizzle. My partner was a hairy yet prematurely bald man named Chris. As he was also big and stocky, the guys at the station had nicknamed him the "Missing Link," but despite the unfortunate handle he was a kind and competent paramedic. We hadn't worked together for a few months, so I was looking forward to catching up.

First I asked him whether he wanted me to drive or attend.

"Today I'm aiming and you're maiming. After we finish checking out the car, let's see if we can grab a coffee before they give us a call."

"Just what I need!"

Chris did the vehicle check while I tested the medical equipment and restocked the cupboards. Then we went 10-8 and headed to the closest restaurant, a small Italian place, arriving just as the owner was flipping the sign from Closed to Open.

"Perfect timing," Chris said as he turned off the motor.

I went inside, bought two coffees and brought them back out to the ambulance.

"It's pretty quiet this morning," he reported. "ALS is running around, but Mother Dispatch hasn't called us yet."

Just then the radio interrupted with the dispatcher's voice: "39 Alpha, cancel—39, cancel."

A voice replied: "Aw, darn, gee whiz, 10-4."

Chris said, "Is that Jeff on 39 Alpha? He sounds disappointed. He must love his work."

"He does, but I also think he wants a call so he has an excuse to talk with that red-headed ER nurse at the General. You know, the tall one with freckles ..."

Just then the radio came to life again: "13 Bravo, 13 Bravo!"

Chris picked up the mike and answered while I wrote down the information. It was a routine call for a respiratory infection, and in a few minutes we were carrying our stretcher, jump bag and oxygen bottle into an apartment building. Upstairs, a caregiver led us to a bed where a wrinkled old man lay propped up on a pile of pillows. His skin was pale and he looked exhausted.

I leaned over him. "Hi there. How are you feeling?"

"Lousy. All week I've been so sick I couldn't walk."

Chris said cheerfully, "If you couldn't walk, I guess you 'flu.'"

The old man looked confused. "Huh?"

The caregiver said, "Mr. Jones is usually pretty active. He turned a hundred and one in January."

"Wow!" I was impressed. "What's it like being a hundred and one?"

"The first hundred years is great. After that, forget it."

I took his vitals while Chris filled out the ambulance form with his address, medical history and medications. The patient was short of breath with a rapid pulse and an elevated temperature; his chest was congested and he was coughing up brown sputum. These are signs of a bad chest infection—possibly pneumonia. In an elderly person with a weak immune system, this could be fatal.

We put an oxygen mask on him, wrapped him up snugly on the stretcher and took him to emergency. The waiting room was crowded, so when I walked over to Lisa, the triage nurse, I said, "We should go first—our patient's pretty sick."

"So? In triage we have two criteria: Are you smiling or not? If you can smile, there's no hurry. If you can't, there's no hope." Lisa looked down at our form. "Of course he's sick," she said. "He's really old. He'll have to wait like everyone else."

"You give priority to the young and the worthless?" I said. "Chris, take him inside!" Chris nodded and starting pushing the stretcher into the emergency room. "There, you can't stop us now."

Lisa wasn't fazed. "Yes, I can. We don't have any empty beds."

Chris stopped halfway through the automatic door to emerg and looked inside. "Maybe you do—the porters are taking somebody away."

Lisa walked over and stuck her head through the door. "Hey, Jo, what happened to your diabetic patient?"

"She has V-tach [ventricular tachycardia] so she's being moved up to 'expensive care.'"

Lisa turned back to us. "You two are the luckiest farts!"

"It's not luck, it's skill," Chris told her.

"Whatever. You win bed four."

The old man was pleased to get in so quickly. "All the staff here are very nice."

"It's the long hours," I said. "It makes them giddy."

We lifted him onto the hospital gurney and I gave my report to Jo. She patted the old man on the shoulder and smiled. "First we have to get you changed. Do you like our designer hospital gowns? They're very fashionable: strapless, backless, hopeless and useless."

We went back outside emergency, and I chatted with Lisa at triage while Chris called dispatch to let them know we were available.

"Are you thinking of moving to the new hospital?" I asked her. "Their ER has better equipment."

She looked disgusted. "Not on your life. The nurses over there are a bunch of cloven-hoofed witches. They ought to install a broom rack in the staff room."

Chris came back waving a piece of paper. "Another routine call. Abdominal pain this time."

We lifted the stretcher back in the ambulance, booked on air and headed off into the rain.

"Aren't you getting sick of sick people?" he asked me. "That's all I seem to do in this town. A good car accident or even a little stabbing would make a pleasant change."

"Enjoy the peace and quiet while you can," I said cheerfully. "We'll have blood and guts soon enough. Right now I'm having fun—I don't even mind the weather. I feel like I'm in a big soap opera full of wonderful, crazy characters. I can't wait to see what happens next!"

Domestic Disputes

People get stabbed and shot in Canada—I've seen my share—but fortunately this doesn't occur too often. I once asked an Oak Bay cop, "Do you see a lot of murders?" He looked puzzled and turned to his partner, saying, "When did we last have a murder—was it ten years ago?"

South of the border it's a different world. One of the paramedics I worked with, Marv, got his Advanced Life Support training in Los Angeles. He and his partner picked up a gangster who had been wounded in a gunfight and rushed him to hospital. No sooner had they lifted him onto an emergency room bed than two members of the rival gang burst through the doors, intent on finishing him off. But the minute they drew their revolvers, an armed guard shot one of them, while an emergency room physician pulled a pistol out of an ankle holster and dropped the other.

To the best of my knowledge Canadian doctors don't pack weapons. Nevertheless, our country is far from peaceful. Our culture is also saturated with violence, and the victims are often women, especially aboriginal women. I have many sad memories. One is of a body leaning against the front door of an expensive West Van house: the man had been killed by his teenaged son during a drunken argument over a new hunting rifle; another is of a terrified woman whose partner had viciously beaten her with a baseball bat.

These experiences can make one a bit paranoid. We were called for an elderly woman who had drowned while kneeling next to her bathtub. She had a deep laceration on her forehead, and a trail of blood led from her bed to the bathroom. It was all very odd, and my partner and I immediately suspected that the husband might have knocked her unconscious, dragged her to the bathtub and held her head under water. We were discussing this with the cop when I realized what had probably happened. While getting out of bed to go to the bathroom, she had slipped and hit her head on the bedside table. Then she had run a full bath to wash herself off, only to faint, tip forward and drown in the water. What the hell were we doing? It wasn't our job to stand around playing Poirot while the poor woman's husband and son were crying in the next room.

Violence is usually ugly, but sometimes it's almost ridiculous ...

TILL DEATH DO US PART

One of our crews got sent out for an assault case. As they walked into the apartment, they saw an elderly couple sitting at the kitchen table. The wife had blood in her hair and an ugly laceration on her forehead that had obviously been caused by having an empty wine bottle smashed over her head. Bits of glass lay around her on the floor.

The attendant who examined her said, "You had better come with us to hospital if you don't want to be left with an ugly scar on your face."

"Nah, I'm not going," she said. "I'm ugly already, so what difference does it make?"

The paramedic tried again. "Well, we're not sure it's safe for you to stay here. Your husband could hit you again. Maybe you should come with us and lay charges."

"Why should I lay charges? I already got even." And with a chuckle she brought her hand up from under the table. In it was a large hammer. "Have a look at the bump on his head!"

Startled, the crew members took a closer look at the husband. Sure enough, an enormous swelling had appeared under the white hair. His face was ashen and he had the dazed look that goes with concussion.

"Are you okay? We should take you to the hospital."

"No. Forget it. I'll be all right."

"She could have killed you! Perhaps you should charge her."

The husband laughed. "Why? I'm not mad at her. I wouldn't have married her if she didn't have spunk!"

BURYING THE HATCHET

Talking about spunky families, one evening a crew was called out to a First Nations reserve for a head injury. The door was opened by a big man in his twenties. The large room behind him was crowded with people drinking, laughing and shouting.

"That was quick," he said. "Thanks for coming."

The attendant asked, "Does someone here have a head injury?"

"Yeah, me," said the man, turning his head sideways. A hatchet was protruding from the back of his skull.

"Good God!" the paramedic exclaimed. "How did that happen?"

"Oh, me and my brother had a little argument. But it's all settled now." He pointed across the room to an equally big man with a ponytail. His brother grinned and waved his beer.

"Look," the attendant said, "don't make any sudden moves. You have to sit down really carefully so we can stabilize your head and that axe. I'm sure the emergency room doctors will be able to safely remove it."

"What?" the man said. "Why don't you guys just pull it out and bandage me up? I don't want to go to hospital. The party's just starting!"

It's a Wrap

The ambulance station was next to a training hospital, which was full of student nurses, and Eddie took full advantage of this, hustling nurses on and off shift. As he was a good-looking, charming paramedic, he was pretty successful. One day he persuaded Jane, a sweet (but gullible) student to meet him after work at the ambulance station. Since the night-shift crews were still in the station, Eddie took Jane through a side door into the garage and then into an ambulance.

"Don't worry," he said, sitting her down on the stretcher. "No one will bother us here. This is a spare car." He opened a bottle of wine, and within twenty minutes they had started to get passionate, their clothes came off and they lay down on the stretcher.

Just then the front doors of the ambulance opened and Robin and Amanda climbed in. They had been given a Code 3 and gone to their car, only to discover that the engine wouldn't start. Fortunately, their station had this spare car. Robin turned the key and pressed the remote, which opened the bay door, then turned on the emergency lights and stepped on the gas. A block away from the hospital he switched on the sirens.

In the back Jane was panicking. Initially frozen with surprise, she grabbed up her clothes from the floor in a futile attempt to cover up and whispered, "What do we do now?"

"I'll get them to take us back," Eddie murmured. Then, louder, he said, "Uh ... guys? I'm afraid you've got some passengers."

Amanda spun around in the passenger's seat while Robin pulled over to the curb and slammed on the brakes. Amanda stared at the tangle of bare flesh. "What the hell's going on?"

"Uh, nothing much," Eddie said. "Just a friendly get-together. But we don't want to get in your way, so would you mind running us back to the station?"

"Get out!" Robin yelled. "We're on a Code 3!"

"Do you mind if we put our clothes on first?" Jane asked nervously. She could see through the side window that they had stopped in front of a busy mall.

"No! Just get out!" Robin roared. "Grab your stuff and get out."

"One second," Amanda said, intervening. "The two of you can take one blanket. Now go!"

So Eddie and Jane found themselves standing naked on a main street sharing one grey blanket, with their clothes, shoes and an empty wine bottle lying at their feet.

I'm guessing it was the last time Jane fell for a fast-talking paramedic.

Part III: The BC Interior

I regularly checked the notice boards for new job openings, and one day an intriguing posting appeared. A unit chief was needed to manage the Ambulance Service in a small community in the middle of British Columbia. Although I'd never even heard of the town before, I decided to apply as it was an opportunity to advance my career. I didn't think I had much of a chance of getting the job, but if I did, I was sure that I would enjoy it. After all, some of the happiest days of my life had been in the small town outside Ottawa where I'd grown up.

To my surprise only three people showed up for the competition; apparently that particular town was not high on most people's bucket list. Each of us was given a pile of cards listing potential problems that a unit chief might encounter, problems such as what you would do if a bomb was placed in the local high school. Or what if your ambulances couldn't get to an emergency because of a snowstorm or your area was hit by a highly contagious epidemic? Or what if you discovered that your station was running over budget or two of your staff members got into a fist fight?

We had thirty minutes to read the problems and write down our solutions. I made it through only half the pile before the time was up, and I went home convinced I had blown the test. I must have done better than my competitors, however, as I was soon issued a pair of striped shoulder tabs to add to my uniform, and a month later Karin and I left the coast and drove northeast to cowboy country.

Winning My Stripes

I soon discovered why few people had applied for the job. The Provincial Ambulance Service had rented space in the local volunteer fire station, which was located in an old three-bay garage. The ambulance sat next to a tanker truck, a dilapidated machine that leaked water steadily all year round, and my desk and the shelves containing our supplies were crowded between an outside wall and the ambulance. The building had no windows, no space for training or extra equipment, and it lacked both heating and air conditioning. In fact, the water leaking from the truck froze in the winter, so when I came in each morning, I had to chip the ice out from under my desk. Then I would sit in my hat, coat and boots in front of a small space heater and do my paperwork.

If I wasn't impressed by the ambulance station, I was even less impressed by the part-time crew. Just after my wife and I checked into a motel, I phoned the number I had been given to arrange to meet my new staff. The next evening I knocked on a door and was ushered into a large living room where ten mostly middle-aged women and men sat on an assortment of couches and chairs.

Everyone looked me up and down silently. Then one of the women said, "So just who the fuck are you?"

Well, I thought, I've been here before! It's going to be sink or swim, just like the logging camp.

I discovered that although all ten of these part-timers could drive the ambulance, only four of them had even basic first-aid skills, and a few were too out of shape to pick up a stretcher. At calls they relied on neighbours or passersby to get patients into the back of the ambulance. Fran, the portly fifty-five-year-old woman who first spoke to me, drove at only one speed: reckless. Even if we were only dealing with a fractured ankle, she would race along the mountain road to the city, careening from one sharp turn to the next, heedless of my pleas to slow down.

But if the local Ambulance Service was a long way from professional, who was I to complain? The pay was so poor that they were really volunteers, doing a thankless job for the good of their community. And I had to hand it to them: after the previous unit chief had suddenly left town two months earlier because of health problems, they had kept the ambulance running twenty-four hours a day.

Clearly I had a lot of work to do. Although we really needed a new station, my first order of business was recruiting more staff and training them to higher standards. But if I was going to train them, I would need better qualifications myself, so I began driving to the nearest city once a week to get certified as a first-aid and CPR instructor.

A Real Professional

At this time I also upgraded my WCB first-aid ticket. At the end of the course each student was examined on a different problem. I was given a photograph of a worker who had been injured in an explosion and was now suffering shortness of breath and a badly broken leg. Well, this is easy! I thought. A blast lung injury with a compound femur fracture. I'd actually had a similar problem a week earlier: a welder had failed to properly wash out a gas tank, and it blew up when he cut into it with his torch. Amazingly enough, even though he was blown six metres into the air and bounced off a bulldozer blade, he'd had only a broken arm.

One of my classmates pretended to be the injured worker, and I started by examining him, then gave him oxygen and finally applied a traction splint to his right leg. Unfortunately my patient didn't look very happy, and while I checked the tension and put ice on the wound, he squirmed and wiggled his left foot.

"What's the matter?" I asked. "Is it too tight?"

"No," he whispered, nodding at the photograph, "you've splinted the wrong leg."

Oh my God! I said to myself. I've just failed the exam! What do I do now?

I looked around to see whether I had time to switch the splint to the other leg. But no chance! The examiner was approaching my side of the classroom, so to distract him, I started telling a joke in a loud voice.

The examiner walked up, listened as I finished the joke and burst out laughing. Then he pointed to my patient and asked, "What type of injury does he have?"

"A blast lung injury with a compound femur fracture," I replied. "Pretty straightforward." I gestured at the photograph, which I'd half-hidden under my patient's pillow.

The examiner leaned down and checked the splinting. "Excellent, excellent," he muttered. Then he turned to the class. "Now see how a real professional does it. Not only does he do a perfect job, but he's so confident that he can tell a joke at the same time!"

Political Blackmail

Although over time I had begun to recruit and train more (and younger) part-time staff, I wasn't having any luck with getting a new station. I sent my regional manager numerous photos of our abysmal facilities, along with endless memos explaining why we needed space for training and room to park a second ambulance. He was sympathetic, but he told me that many small towns were in a similar predicament, so we would just have to join the queue. Moreover, no matter how deserving we were, nothing could be done because the provincial government had put a budgetary freeze on all new construction: not a single hospital, clinic or school was being built anywhere in British Columbia.

The situation seemed hopeless, but I had forgotten the power of our two oldest part-timers, Fran and Ginny, who happened to be the chief organizers for the local branch of the political party currently in power in Victoria. And an election was coming up. When their local member of the legislative assembly—who was also a top provincial government minister—came to town, they requested a private meeting.

As soon as they were alone, Ginny said, "Minister, this town needs a new ambulance station. The one we have to work out of is a dirty disgrace."

The minister shrugged. "I'd love to be able to help you, but you know that our election campaign is based on fiscal responsibility. We have publicly pledged not to construct another government building for at least two years. Come back to me after that, and I'll see what I can do."

"No," Fran said. "You're going to build us a new ambulance station now. This is a swing riding, and you need the votes from this town or you won't get elected, and if you don't get elected, your government might lose its majority and get kicked out of power. So here's the deal. You get us the station and we will support you, like always. But if you don't come through, we'll tell everyone in the entire district that you're a useless fraud and convince them never to vote for you again."

Two weeks later the government budgeted $450,000 to build a brand-new ambulance station in our town, the only new government building approved for construction anywhere in British Columbia, and it was a beauty. The regional manager called me as soon as he heard the news. "I don't get it," he said. "Why you? Why now?"

"Just lucky, I guess," I said, but luck had nothing to do with it.

Lost in Translation

The call to the Kamloops dispatch centre came from a prosperous and well-organized First Nations reserve. Because it was a small community, even though there were no street signs or numbers on the houses, the people who lived there had no problem finding their way around as every family's name was carved on a post at the foot of their driveway. However, it was easy for outsiders to get lost, especially since many families shared the same last name, and this situation caused the occasional mix-up.

"My brother has broken his leg," the caller said. "We've got him on the couch at my house. Could you send up some help pretty quick?"

"Sure, we'll have a crew on the way in a minute," the dispatcher said. Knowing about the lack of street addresses on the reserve, he added, "But how do we get there?"

"How do you get here? Don't you guys still drive those white trucks with the red stripes?"

Country Calls

Ambulance calls in the country are often quite different from those in the city. Highway accidents can be worse because they tend to take place at higher speeds, hasty hunters shoot themselves or each other, and cowboys suffer injuries from horses or angry rodeo bulls kicking them or tossing them over fences. And at times the job can get quite smelly. We were called to an accident on a back road involving a truck full of teenaged army cadets. The driver had been going too fast, and at a sharp bend the truck had slid on the gravel and rolled over in the ditch. The good news was that the cadets had all been thrown onto a huge pile of manure and no one had serious injuries. You can guess the bad news.

CHAINED

By this time Karin and I had bought a house on the outskirts of town, a pretty location surrounded by cattle ranches on three sides; during the day horses leaned over our fences, and at night we could hear coyotes yipping. It wasn't long before I had convinced Jesse, the farmer across the road, to join our ambulance crew, along with Angie, who lived on a nearby ranch. Soon they were joined by Jed, a wiry young rodeo rider who always wore faded jeans held up by trophy belt buckles. I enjoyed working with them as they were conscientious and very keen. One December when we got an emergency call in the middle of a snowstorm, I knew Jed had arrived at the station when I heard a loud crash. When his pager went off, he had raced down the road, hit the brakes on a hidden patch of ice outside the station and skidded into a telephone pole. Fortunately his pickup truck was already so banged up that a few more dents didn't make much difference.

I soon had twenty part-time staff, all trained to at least minimal emergency medical and driving standards. Our station also raised funds for extra equipment, and with the generous support of the community we were able to buy and equip a rescue truck. But of course life is never perfect. Now my big problem was how to keep two ambulances manned around the clock in such a small town—most of the people on my roster had other priorities, including working and raising kids. And they had lots of other demands on their time: Christmas, summer holidays, curling, hunting, baseball, fishing, visits from the in-laws. I discovered that it was easy to fill the schedule in March, but the rest of the year was tricky, and nine out of every ten days I would have to put my own name down to work on either our first or second car.

One February I was pleased to have a weekend free, especially as it was snowing heavily. I turned off my pager and spent a pleasant time at home with my family—which now included two children. This lasted until Sunday at 10:00 p.m. when the telephone rang.

It was Joe, the dispatch supervisor, and he was so mad he was stuttering. "Gr-Gr-Graeme!" he shouted. "Is everyone in your station a total m-moron?"

"Why? What's the problem?"

"I j-just got a call from the dispatch centre that your crews have wrecked both of your ambulances. T-two cars at the same time!"

"Is anyone hurt?" I asked. "What happened?"

"No, nobody's hurt," he grunted. "We called out your first crew for a Code 3 MVA, and because it's snowing they took the ambulance you keep chained up and raced down the highway. Of course the chains expanded and one of them wrapped around the muffler and put the ambulance in the ditch! Do your part-timers come from the tropics? Don't they know that you can't drive with snow chains at high speeds?"

"Well," I said, "everybody makes mistakes from time to time. What happened to the other car?"

"The dispatcher called out your second crew and sent them to the MVA, and then sent a tow truck to get your first crew and their ambulance. So your second crew took a couple of patients to the hospital, and on the way back they stopped for gas, and that nitwit Mary—is that her name?—filled the tank with diesel fuel. Fortunately she realized her mistake and called dispatch before she started the engine. How dumb can you get?"

"Now, now," I said, "don't be too hard on Mary. She's a trained nurse, so she's usually a real professional."

"She's a professional idiot!" Joe snarled. "I was off work enjoying a hockey game, and now I have to deal with this mess!"

"I guess that's why they pay you the big bucks," I said. "I'd appreciate it if you could dig up a spare car for us. Mary and her partner can switch over the equipment and bring it back with them. Thanks for your help."

"Right," Joe muttered sourly. "I'll see if I can repay the favour some time."

After I hung up, I laughed. A few years earlier I would have been embarrassed to have my staff make stupid mistakes, but now I was confident that, despite the occasional glitch, our station was running smoothly. Nevertheless, I figured I had better teach another class on emergency driving.

FLIPPED

Paramedics see patients for only a short time and rarely get much feedback. Although patients occasionally send letters of thanks, in my experience it's the little old ladies who are more likely to write letters after they have been helped back into their wheelchairs than middle-aged men who have been brought back to life.

But there are pleasant exceptions. I was standing with a couple of our part-timers outside our station, watching the pilot of a light plane practising landings on the local airstrip. Suddenly a gust of wind caught the plane and it flipped over, crashing nose down next to a schoolyard. I immediately notified the dispatch centre, and in a couple of minutes we were at the scene. We were afraid the leaking fuel might be ignited by an electrical spark, so even though the pilot had broken limbs, we pulled him quickly out of the cockpit. After that rough treatment I didn't expect any praise and was surprised when our patient—who turned out to be a medical doctor—sent us a gracious letter of thanks.

SWAMPED

But there are times when we feel more used than appreciated. Two of our part-time crew responded to an accident where a car had plunged over a steep embankment into a swamp. Fortunately, the vehicle landed wheels down on some bushes, and the driver didn't even get her shoes wet. But the paramedics had to wade chest-deep through the swamp to reach the car. When they got there, the driver, an extremely large woman, told them that she was unable to walk because her back hurt. So they put her on a stretcher and carried her across the swamp and up to the ambulance. Once on dry land she stood up and said, "Thanks, boys, I'm feeling much better now." Then the woman flagged down a passing car and disappeared. The crew, exhausted and dripping with mud, looked at each other with chagrin. She had really taken them for a ride.

Bottoms Up

The nursing home was located in a beautiful country setting, but it had the drawback of being forty-five minutes away from the nearest hospital. So when a resident had a nasty fall around ten one evening, the nursing staff were reluctant to take him all the way into town.

Instead, a nurse called the local GP. But there was a problem: he was off duty and didn't want to see anyone. The nurse insisted that their resident would only require a couple of stitches. "All you have to do is go downstairs from your living quarters to your clinic and sew up the cut in his left eyebrow. It won't take you more than a few minutes."

However, when they arrived, they realized that the doctor had been drinking. They had known he had an alcohol problem, though it usually didn't affect his work, but now his voice was slurred and he swayed when he walked. He was plastered!

The doctor sat beside the patient, leaned over him and proceeded to stitch up the eyebrow, and—even drunk—his stitches appeared neat and professional. Unfortunately, the nurse didn't get a close look at his handiwork until the job was finished. When the doctor stepped back, she saw with horror that he had not only sutured the cut in the old man's eyebrow, but he had also sewn his upper eyelid to the eyebrow!

What should she do? The doctor was in no state to fix his mistake.

Diplomatically she thanked him and saw him safely back up the stairs to his house. Then she reached for the telephone and called an ambulance. The crew took the old man into emergency, where his eyelid was cut free and the sutures redone. The whole incident was covered up as no one wanted to ruin the doctor's reputation. After all, he was off duty at the time and had tried to refuse the job. And the patient didn't complain. Sadly, he had Alzheimer's, but perhaps not so sadly, he couldn't remember a thing.

Joe's Place

As soon as the console lit up, the Kamloops dispatcher could see that the phone call was coming from Lillooet.

"Someone's been stabbed!" a voice yelled in his ear. "Quick, send an ambulance!"

"Could you tell me where the stabbing has occurred?"

"Where do you think? At Joe's place. Now hurry up!"

"Excuse me, sir, but where is Joe's place?"

"You don't know where Joe's place is? Everybody knows Joe's place. Just send the ambulance!"

"Okay, sir, but please calm down. I need to know more about the injury. Exactly where was the person stabbed?"

"Are you deaf or something? How many times do I have to tell you? He was stabbed at Joe's place!" *Click!* The caller hung up.

The dispatcher paged the Lillooet crew. Thirty seconds later they called in. "What's up?"

"All I've got is that some guy has been stabbed at Joe's place," the dispatcher told them. "But I'm afraid I don't know where it is."

"What? You don't know where Joe's place is? Everybody knows where Joe's place is."

"Well, good. Go there, Code 3."

"Okay, we're out of here. But first, just where exactly was this guy stabbed?"

The dispatcher was waiting for this. "Are you deaf or something! How many times do I have to tell you? He was stabbed at Joe's place!" *Click!*

The Mounties

I spent five years as the ambulance station chief in that small town. During that time a permanent force of three Royal Canadian Mounted Police officers was also stationed there. They worked out of a modern, bullet-proof fortress, which was overbuilt according to lengthy bilingual specifications formulated in Ottawa: it probably cost as much as half of all the houses in town put together. In any case, they didn't have much to fear from the local inhabitants, who showed little inclination to get off their stools in the Legion Hall or the chairs in the bingo parlour to mount a violent assault on the Queen's representatives. The worst offenders in town tended to be drunken cowboys, pot-smoking teenagers and trigger-happy hunters. In fact, the bush was so dangerous at the beginning of the season that one of my friends suggested that the easiest way to bump off enemies without getting caught would be to glue some twigs on their heads and tie them to trees; eager hunters would finish them off in no time.

SPILT MILK

Every year the town held a rodeo, and inevitably one of the organizers proposed a competition between the three local cops and three local paramedics to see who could be the first to catch a cow, milk it and bring a full bottle back to the judges. We were sure it would be a slam dunk as we had a professional cowboy and a third-generation rancher and dairy farmer on our side (plus me) and the Mounties' team comprised two city boys plus one fellow from a family of Manitoba wheat farmers.

When the time came for our event at the rodeo, we stood confidently on the side of the arena beside the three big cops. They had nothing in their hands apart from the empty milk bottle held by Len, their ex–farm boy. On our team, cowboy Jeb stood expertly twirling a lasso, while farmer Jesse held our bottle. Then two cows were pushed into the arena, a judge fired a starter gun and we ran after our cow. Jeb threw his lasso, and I was certain that the game would be over in a minute. But the rope bounced off the side of the cow, and the poor beast, thoroughly terrified by the sight of three charging men, took off at a gallop.

Out of the corner of my eye, I noticed that the cops hadn't been running. Well trained in crowd control, they had spread out and were walking slowly and methodically toward their cow, which edged nervously away from the centre of the ring. In a few moments they had her pinned against the fence, where Len quickly filled his bottle. Cheered by the spectators, they raced back to the judges' stand.

In the meantime, we were still chasing our cow around the arena. Although Jeb kept throwing the lasso, the cow dodged and jumped, and the rope wouldn't hold. Finally, the rope caught on a back leg, and Jeb pulled one way while I rushed in and grabbed the animal around the neck. Jesse quickly knelt in the dirt and started milking, but the cow was bucking so hard that I had trouble keeping my grip, especially as I was afraid that a horn would take out my eye. That's when the cow broke free again and kicked the bottle, covering Jesse with milk.

The onlookers roared with laughter. Humiliated, I threw caution to the wind and jumped on the cow, while Jesse dove between the flailing hooves and desperately tried to get some milk. It took us another five minutes, but eventually we were able to complete our task and limp back to the judges. The whole time the crowd laughed, hooted and jeered.

That was the first and last time I entered a rodeo. They say you shouldn't cry over spilt milk, but believe me, that day I almost did.

GOOD COP/BAD COP

The local RCMP detachment was headed by a big, ham-fisted corporal named Jock. At the time I didn't have a high opinion of him because one of our paramedic part-timers had once phoned him around midnight to report a herd of cattle wandering down a country road. "Maybe you should get them off the road before someone has an accident," the paramedic said.

Jock was furious. "Did you wake me up because some cows got loose? What do you think I am—a cowboy? That's the farmer's problem, not mine!" And he hung up.

After that incident I wrote him off as incompetent and lazy although, to be honest, we never had problems when we worked together. On one occasion, we were both sent to an accident where a pickup truck had skidded out of control and flipped upside down. Its doors were jammed shut, three unharmed but panicking teenagers were trapped inside and broken glass was strewn on the ground.

Rather than taking the time to pry open the doors, I got down on my back, swept the glass out of the way with my jacket sleeve, wormed my way under the overturned box and got the kids out through the cab's back window. Jock was impressed. "That was brave!" he said. I didn't really think so because the wreck was stable, but I wasn't about to argue with a good opinion.

Another time our ambulance was sent to the local bar on a Friday night to deal with a young guy who had gotten completely tanked and walked into a swinging door. Even though he had lacerated his forehead and the blood had run all over his face and down the front of his shirt, it was a minor injury, and as soon as we loaded him into the ambulance and wrapped a bandage around his head, he fell fast asleep. We were about to drive off to the hospital to get him stitched up when his sloshed girlfriend opened the side door and stumbled in.

"Oh no, he's dying!" she shrieked, throwing herself on top of him. "Don't die, Wesley! You can't die on me, I love you too much!" She wrapped her arms around his neck, knocking off the bandage, and the cut started to bleed again.

"Okay, that's enough!" I said, trying to pull her off the stretcher. "Say goodbye—we're taking your friend to the hospital to get stitched up. You have to let go."

"No!" she cried, tightening her grip. "I can't leave him. He's dying! He needs me!"

I turned to my partner Brad. "We'll have to throw her out—but gently. We don't need another patient." Brad took one arm, I took the other, and we pried her off her boyfriend and lifted her up in the air and out the side door.

By the time we deposited her, kicking and screaming, on the ground, a crowd of twenty or thirty drunks had come out of the bar to see what was going on. The girl waved to them and yelled, "Help! Wesley's dying! He's dying and they don't care!" In response the drunks surged toward us.

"Oh-oh," Brad muttered. "This doesn't look good." Quickly we shut and locked the ambulance doors, and Brad climbed into the driver's seat, started the engine and turned on the lights and sirens. But now the crowd had us surrounded, and we couldn't move. Then some of the younger men started to pound on the sides and rock the vehicle.

Fortunately, the 911 dispatcher had called the Mounties at the same time he called us, and at this moment Jock showed up. He was alone, but he pushed his way into the crowd and started shoving troublemakers away from the ambulance. For a moment the path ahead was clear, and Brad pulled away. As we left, I looked back and saw Jock standing in the middle of that angry mob. Although he was a big man, he was badly outnumbered. But he had guts, he was wearing his uniform and he wasn't backing down.

With time I've learned that one shouldn't make assumptions, and, thinking back on it, maybe Jock wasn't such a bad policeman after all.

THE MAN HUNTER

The three Mounties in the detachment had very different personalities. Whereas Jock could be bullheaded, Len was outgoing and friendly, and Francis was distant and reserved. I had little to do with Francis, though I would occasionally spot him stopping cars or setting speed traps. But of the three, he would have the biggest impact on my life—along with a man called Will.

Will, who must have been in his mid-fifties at the time, had moved to town after he found a property there that was suitable for training guard dogs. He built a big compound filled with dog pens and obstacle courses and surrounded it with a high fence topped with barbed wire. Then he hired June, a blond, athletic twenty-five-year-old, as his assistant.

Although June loved working with big dogs, she was also interested in becoming a paramedic, so she dropped by the station to apply for work. I was happy to hire her because I needed to maintain my roster of part-time staff to keep our two cars crewed around the clock. After that, she came over whenever she was free, and I spent a fair bit of time instructing her in emergency driving and other ambulance skills.

For a few months everything went smoothly. Then one afternoon she rushed into my office with a frightened look on her face. "Graeme, you've got to help me! Will told me that he wants me to be his partner—not only his business partner but to live with him. When I said I wasn't interested in that kind of arrangement, he said I didn't have a choice, that we've already been together for almost a year and I can't back out now. And he says because that's what he wants, that's what's going to happen."

"You'll just have to be firm with him," I said. "Of course, if he doesn't change his tune, you may have to quit your job."

"The problem is, I'm frightened of him. He told me he used to work for the CIA in Vietnam and Africa, and that's where he learned to handle dogs—he used them to track guerrillas in the jungle and find hidden tunnels. But he says he wasn't just a man hunter, he was also an assassin. He was taught all kinds of ways to kill people. I believe him—he keeps the ears of people he's killed on a string. He even said that he's gotten rid of people who crossed him here in BC by tampering with their car brakes. I'm scared! How do I know he won't murder me if I try to leave?"

I asked, "Has he hit you or threatened to hurt you?"

"Not yet."

"If he does, you'll need to go to the police. But frankly, I'd suggest you drive straight back to the city. When you get there, give him a friendly call saying that your grandmother has broken her hip in Montreal or New York and you have to look after her. Then just disappear for a while and never contact him again."

She pondered the suggestion for a minute. "Okay, that's what I'm going to do. Thanks, Graeme."

I wasn't happy to lose a keen employee, but *c'est la vie.* Two days later I dropped by the hotel bar looking for Byron, one of our part-timers. I had made it a policy never to go into a bar while in uniform, but I desperately needed to plug a hole in the roster, and Byron's wife had told me that he'd finished work early and gone to the hotel for a drink.

Byron and his friends were at the far end of the bar. I talked him into signing up for the vacant shift and was walking toward the door when a voice called out, "Graeme, come here! I need to talk to you." It was Will. He was sitting by himself at a table covered with empty beer glasses.

"Hi Will. What's up?" I said as I walked toward him.

"I know you've been fucking June. I'm going to get you."

I was taken by surprise. "What? Nothing is going on between us. She's an employee."

I thought he might have had too many drinks, but his gaze was steady and his voice clear and cold. "I know you're fucking her. I'll get even. I'm going to get you."

I was shaken, but I tried not to show it. "You're nuts," I said and walked out of the bar. Now what was I supposed to do? June could leave town but I couldn't. If he really was a professional killer, he could ambush me anywhere. And I had a wife and two kids. He could always get to me through my family. That night I borrowed a rifle and kept it loaded beside our bed, but I didn't get much sleep.

The next morning I went to the cop shop and explained the situation to Len. Then I went looking for Will. I found him drinking coffee in his neighbour's kitchen. Evan was the local beekeeper, and my wife and I were not only his regular customers but we had become good friends with him and his young family. When he saw me driving up, he opened his front door and waved me in. "Welcome, Graeme. Would you like a coffee?"

"No thanks, Evan. I'm just here to say something to Will."

I sat and looked Will in the eye. "Will, I've told the cops and my brothers that you said you were out to get me, so if anything happens to me, they'll pin it on you."

Evan's mouth dropped open, but Will's expression didn't change. "I hear you," he said. "I hear you."

I got up and left, not sure of the next move. But two days later Len searched Will's compound, confiscated a sawed-off shotgun and warned him to stay away from me and my family. I hoped that was the end of that problem. I wanted that psycho out of my life. My goal was to be a paramedic, not a paranoiac.

A TWIST OF FATE

Meanwhile, June had disappeared and things soon settled back to normal. Until September, that is, when our friend Emma called. A year earlier her pregnancy had ended in a stillbirth, and she had fallen into a deep depression. She'd quit her job, and because her partner couldn't cope with her grief and anger, their marriage had collapsed. She was getting counselling, but it wasn't helping, and now she was feeling suicidal.

My wife and I suggested that she come and stay with us for a while. Emma loved our young children, and we were sure it would be good for her to move far away from her apartment with all its memories and live with us in the country. She agreed, and a week later Karin and I picked her up from the bus depot.

It was just what Emma needed. She enjoyed playing with the children and the dog and started to help around the house, even doing some baking. Her mood steadily improved, and after a couple of months I became confident enough in her caretaking skills to suggest to Karin that this might be a good time for her to take a brief holiday. Because I was always on call, we were rarely out of town, and Karin had almost never spent a night away from the children. Emma said she felt ready to handle a little responsibility, and my wife left for a five-day visit with friends in Vancouver.

At first all seemed well. But on the third day Emma started to feel anxious. "Please don't leave me alone, Graeme. I'm afraid that I'll start obsessing about my baby and fall back into depression."

"I'll spend as much time with you as I can," I replied, "but if I get called out, I'll have to go." Luckily we didn't get a call and I was able to spend the whole day at home.

The next morning she seemed the same—still anxious but able to cope. I told her that I would have to go to my office for at least four hours. "As I'm in charge of the only Ambulance Service in this area, I have to show up at the station once in a while."

"Please don't leave me alone," she repeated. "I don't feel safe."

I tried to reassure her. "If you start feeling tense, give Sue or Amanda a call—they're just down the road. I'll only be gone for a little while and Karin will be back tomorrow. Hang in there! You'll be okay." She didn't look very confident, but I had work to do, so I left.

Three hours later my office phone rang. It was Emma, speaking in a soft, tired voice. "I can't do it anymore, Graeme—I've taken all my pills. Don't worry about the kids—I've left them with your next-door neighbour."

I broke into a cold sweat. "No, Emma! Don't do this! Make yourself throw up and then lie down. I'll be home in five minutes."

"I've made up my mind. I'm going to the bush. I'll find some peaceful place and go to sleep. Bye." She dropped the phone.

I was frantic. I picked up the hotline, called dispatch and explained that my partner and I would be taking the ambulance up to my house because my friend had taken an overdose.

The dispatcher said, "Now stay calm. Go to your house and see whether your friend is still there. If she's left the property, don't start looking for her. I'm sending the Code 5s as backup. They can organize a search party if necessary."

We raced up the hill to my house. Next to the kitchen sink were three empty bottles of antidepressants and pain medications. We took a quick look through the house and were searching the outbuildings when an RCMP cruiser drove up.

Francis got out and I ran over to explain the situation. "There's no sign of her. She said she was going to the bush and the back gate is open, so she's probably gone that way." On the other side of the fence was a big ravine, and beyond that, endless kilometres of scraggly woods. "My partner and I will go that way and start looking," I said.

"No, you won't!" Francis said. "I'm going to organize a search and you're going to follow my orders. Now get back in your ambulance and stay there. You'll be needed as soon as we find her." He went back to his squad car and got on the radio.

The next half hour was one of the worst times in my life. I knew that our young friend had ingested enough drugs to kill her, and every passing moment was bringing her closer to death. The police had put out a call for help, and to my amazement within ten minutes dozens of volunteers started to arrive. As they came, Francis marshalled them into teams and told them where to go. They were soon joined by a search-and-rescue plane flying low over our heads. My hopes began to rise, but they were dashed when the teams returned from their first sweep. No sign of Emma!

Francis had asked the volunteers to do a rapid search in every direction within three hundred metres of my house. Now he asked them to go back and look again. "That's crazy!" I protested. "She's not there. She said she was going into the bush. You need to look farther away from the house."

Francis pointed his finger at me. "I told you to stay in the ambulance.

Most of the time people are found close to their last known position. Now shut up and let me do my job!"

I went back to the ambulance and sat down, my hands shaking. Then, through the windshield, I noticed that Will had turned up with a big German shepherd, and after talking to the Mountie, he joined the search.

The minutes ticked by and my heart sank. I thought it must be all over for my friend and felt guilty for leaving her alone with the kids. Why was this happening? She was too young and too sweet to die like this. Then I heard shouts. It seemed that Emma had climbed into an old, unused water tower and hidden in the bottom. Although the searchers had looked inside during the first sweep, they hadn't seen her in the shadows. But Will's dog had picked up her scent and found her.

They carried Emma to the ambulance. She was unconscious, but the doctor at the local clinic pumped her stomach and she survived. And although it took time, her physical and psychological health fully recovered.

I've made many mistakes in my life, but that was one of the worst. We all have breaking points, and sometimes things are too painful or frightening for us to bear. Emma had warned me that she was losing it, and I hadn't listened. We were lucky that so many other people were there to help.

To show my appreciation, I bought two bottles of twelve-year-old whisky. I gave one to Francis and the other to Will. When Will opened his door, I held out the whisky and said, "Thanks for saving my friend's life." He looked at me without a smile. Then he nodded, took the bottle and shut the door.

I didn't see Will again for another ten years. One day I was waiting in line at a supermarket checkout counter in Victoria when I realized that he was standing in front of me. He paid, and as he left, he turned and saw me. Neither of us spoke a word. Then he walked away.

Unsafe Sex

They were obviously a well-matched couple: both middle-aged, both drunk and both sitting in the same Prince George bar at two in the afternoon. So it wasn't surprising that by 2:45 they were back at her place in her bed, their clothes scattered all over the room.

The surprising part was the reaction of the woman's two teenaged sons when they got home from school a little after three. When they heard strange noises upstairs, they threw open the bedroom door to discover a stranger bouncing around wildly on top of their mother. In a rage they ran back outside to the garage, picked up lengths of pipe and raced back inside, where they began furiously beating the confused stranger as he desperately struggled to get his pants on.

The neighbours, frightened by all the yelling, phoned for help. But by the time an ambulance and a policeman arrived, the man had fled for his life into the bush across the road, vigorously pursued by the woman's righteous sons.

"Well, what do we do now?" one of the paramedics asked the cop. "By the time we find this guy the kids will have killed him."

"That's what my dog is for." The cop proudly let a big German shepherd out of the back of his patrol car. "Go get 'em, Thor!" The dog leaped away eagerly, nose to the ground.

In a few minutes the bush erupted with shouts and barks. The clamour grew closer until a figure burst out of the woods. It wasn't one of the teenaged attackers, however, but the older man, stumbling along on bare feet. As he drew closer, the dog growling and nipping at his heels, they could see that he was a complete mess. His head and naked torso were covered with bruises, and his arms and shredded pants bore the marks of dog bites.

As the cop called off the dog, the man limped up, obviously relieved to have reached the ambulance. "I am safe! Finally, I am safe!" He spoke in a thick accent. "I am so happy that nice dog he found me!"

And wouldn't you know it? The woman had stolen his wallet.

THE COP WHO COULDN'T SMILE

That town was such a peaceful little place that most of the time the Mounties just cruised up and down the highway in search of speeders and coffee shops, crawled through the high weeds on their hands and knees in search of high school students sharing forbidden joints and smooches, or rifled through their files in search of missing forms and circulars. This meant that they were usually available when paramedics needed them to act as escorts.

On one of these occasions our patient was a six-foot-two, powerfully built, paranoid millworker. He sat nervously in the back of the ambulance, jerking his head back and forth and uttering confused threats. We carried restraints, but there was no way I was going to jump on top of him and try to tie him upside down on the stretcher. Moreover, only the police and doctors have the authority to order restraints.

Instead, I just tried to look relaxed and non-threatening as I talked to him quietly from a position close to the side door. If he got violent, my driver and I had decided we were just going to slam on the brakes, get out and run for it. We weren't being paid to slug it out with a lunatic, and he could push the ambulance off the road and down the cliff if he wished. Meanwhile, as I was busy being pleasant to the patient and he was busy being hostile to me, my driver was radioing dispatch to arrange for a Mountie to meet us. We pulled over by the side of the road, and minutes later a patrol car parked beside us and a tall cop climbed on board. I was glad to see that it was Len, the cop I had the best rapport with among the town's Mounties. A big, friendly fellow, he came from a small community on the Prairies, and I was sure he would handle a paranoid patient in a sensitive fashion.

Len, however, was a long way from being relaxed and sensitive that day. He sat down abruptly at the end of the bench and stared at the patient as if he wanted to murder him. "Just lie down quietly and don't move," he muttered, "or you'll wish you were never born." Our patient lay down meekly, and Len turned to my driver and grunted, "Move it!"

The drive to the city took another forty minutes. The whole time Len sat staring at the patient, his face contorted and teeth grinding. His fists clenched and unclenched, and his feet shuffled as he crossed and uncrossed his legs. The patient, completely terrorized, huddled against the wall.

As for me, I sat quietly, shocked by Len's volatile mood. I'd always thought of him as the friendliest policeman I had ever met, and here he was ready to beat our patient to a pulp. I began to wonder whether his usual smile was just a cover for a vicious, psychopathic personality.

When we reached the hospital, Len immediately disappeared. After turning our patient over to the psychiatric staff and hospital security, we put new linen on the stretcher and went back outside to the ambulance. Len was waiting for us. "Can you give me a ride back to my car?" He was cheerful, the old smile back on his face.

"You want to go back with us?" I was surprised and not too sure I wanted his company. "You didn't seem too happy to be with us on the way down. I figured you were furious because we made you waste your time playing escort."

"No, you've got it all wrong," he said and laughed. "I was just heading for the bathroom when you called for help, and it nearly killed me holding it in all the way to the city!"

My Worst Call

It was one of those really hot summer days in the Interior, and I was hoping to spend a cool afternoon indoors in our new station. That dream evaporated when my pager went off. I called in.

"Take a Code 3 to the Last Lake Resort for shortness of breath. It seems one of the owners just had her chest kicked in by a horse."

"Don't you want to send a city car? We're at least forty-five minutes away."

"Sorry, all the cars are tied up."

"Gotcha."

While I waited for my partner, I stared at the poster on the wall announcing the upcoming annual Ambulance Fishing Derby at Last Lake. The derby meant little to me as I don't fish, but I knew that this week of family fishing was a summer high point for many of my co-workers. The owners of the resort knocked themselves out to provide a good time for this heroic and dedicated group of public servants.

When my part-time partner arrived, all freckles under his thick red hair, I warned him, "We've got to move it on this one. If we don't give them a little bit of their good service back, we can kiss goodbye to the fishing derby and our reputations with it."

"Don't worry," Bill replied. "I can make this box fly!" And fly it did, all the way down the winding valley road to the Last Lake turn-off. From there it was a long climb up a twisting gravel road to where the lake was hidden far, far back in the hills. The old ambulance bounced and skidded up the road, and as we charged into each turn, Bill slammed on the brakes then tromped on the accelerator to pull us away again on the straight. The engine whined and stones spattered on the floorboards beneath us.

Suddenly a white cloud engulfed Bill, and he let out a scream. "We're on fire! My foot's burning!" He skidded the ambulance to a stop and leaped out the driver's door. I grabbed the microphone and called dispatch.

"This is 4479. We're on fire and out of service. You'll have to send assistance."

"Ten-four, 4479. Do you need a fire truck? Any injuries?"

"Hang on," I said.

I ran around to the driver's side, where my partner was hopping around, holding his right foot. I looked under the cab. "Oh, for God's sake! There's no fire—the radiator hose has burst." I was mortified. Not only were we broken down, but now I had to tell every crew in Region 4 that happened to be listening in that we were too stupid to tell the difference between fire and water. My sympathy for Bill disappeared.

I got back on the radio. "4479. It's just a burst hose. My partner is a bit scalded but it's nothing serious. He can wait for assistance. I'll head for the resort on foot."

"Ten-four, 4479. We'll send another car along as soon as we can get one clear, but it's going to be a while."

Leaving Bill sitting on the tailgate and moaning softly, I started running up the road with the portable oxygen tank and a jump kit. Soon I was covered with sweat and gasping for breath.

The road went on interminably, up and up, with nothing on either side but empty fields and pine woods. The twenty-five kilos of equipment I was carrying felt like an entire emergency room, and as the sun burned into my neck I began to think I had somehow been entered in a marathon across a tilted Sahara.

Finally, I heard the sound of a truck. With enormous relief, I set down my equipment, stood in the middle of the road and waved. The truck came toward me, blared its horn and gunned on past, leaving me coughing in the thick dust of its wake.

For another twenty minutes I stumbled on, the sweat leaving brown streaks on my uniform. Again I heard an engine. Again I stopped and waved. This time the driver pulled up beside me. I explained that I had to get to the Last Lake Resort quickly. No problem. We were there in five minutes.

I stumbled in the door to find a middle-aged woman sitting bolt upright in the centre of a large room, her husband standing anxiously at her side. She was breathing with quick, short breaths and obviously in much pain.

"I'm sorry I took so long," I stammered. "The ambulance broke down."

"Why did you bother coming at all?" the woman said as she gasped for air.

I tried to be reassuring. "I don't have much with me, but this oxygen will help your breathing and reduce the pain." I turned on the tank and there was a sudden loud pop. Damn! The gasket had blown and the spares were in the ambulance! I'd never had this happen before. "I'm terribly sorry," I said, "but the regulator has just packed it in. But let me examine your injuries."

She gave me a look I will never forget and whispered, "Don't touch me. Just go away."

I sat in a corner until the ALS crew from Kamloops arrived. When Matt walked in, all I could say was, "Matt, please take over. I've got to get out of here."

The owner of the Last Lake Resort probably recovered from her injuries long ago, but let me tell you, I don't think I'll ever get over that call.

On (and Off) the Road

People have always had accidents on the highway, and despite better education and safer technologies, they always will. My ambulance part-timers had their share of close calls. Angie was driving home along a mountain road when a large boulder tumbled down a steep cliff and crashed through the roof of her car. She was unharmed—the rock fell on the rear seat—but her car was a total write-off. Bill—the one who was scalded by the burst radiator hose—had a similar close brush with death. While driving through the Fraser Canyon on a hot, sunny afternoon, he lit a cigarette with a Bic lighter, and then tossed the plastic lighter on top of the dash, where it got stuck in an air vent next to the windshield. The windshield must have acted like a magnifying glass, because the lighter suddenly exploded, blowing out the windshield, burning his hands and temporarily blinding him. Shocked, he slammed on the brakes, fortunately sliding into a nearby retaining wall rather than tumbling down into the river.

A REAL CLIFFHANGER

I didn't think there would be any survivors on this one. The dispatcher said that a car had missed a turn, broken through the safety rail and plunged into the river canyon, and from the location I guessed it was lying at least sixty metres below the road. This was going to be ugly, and both of our crews had been paged. Our two ambulances carried the Jaws of Life and other extrication equipment, and I had also stocked the main car with climbing gear. I threw a basket stretcher into the back along with a couple of extra ropes, pulled the ambulance out of the station and waited. In a few minutes an old pickup truck skidded to a stop. Angie jumped out and I waved her into the driver's seat.

"You drive," I told her. "You know every twist and turn on this road."

Angie had grown up in the Interior, and she could handle the ambulance on the mountain curves like it was a racing car. Nevertheless, it was twenty minutes before we saw half a dozen vehicles lining the road next to a row of broken guardrails. We pulled up behind them and joined the small crowd peering over the edge.

Below us, about halfway down to the river, the roof of a black car sparkled in the sunlight. On its way down, the car had been caught by some trees growing out of the side of the cliff: it was sitting right side up and looked remarkably undamaged.

"Do you know if anyone is alive?" I asked a bystander.

"We heard some voices, but we can't make out what they're saying."

"Angie, I'll go down and assess the situation. When the second car arrives, send someone down with the basket stretcher." I put on my climbing harness and tied on a safety line. I gave the rope to Angie, who anchored herself to an unbroken post. Next, I searched for an anchor for my main rope. The best spot seemed to be the bumper of a big four-wheel drive truck. "Is the owner here?" I called out. No answer. "I'll tie the rope off beside his bumper winch," I told Angie. "Tell the driver not to move his truck until I'm back on top."

I slung the jump kit over my back, clipped on my rope and started rappelling down the cliff. Angie leaned over from above, carefully belaying my safety rope. Because the rock face was almost vertical, it was an easy, quick descent. The bystanders hung over the edge, watching intently.

Suddenly I started going back up the cliff. "Whoa! Stop!" I yelled. Above me the heads disappeared and I heard confused shouting. The rope stopped pulling me up. Then I was slowly lowered back down, and Angie and the bystanders reappeared. "It's okay," she called. "The truck's back in position." I found out later that the driver of the four-by-four had decided to leave, not realizing that my rope was tied to his bumper. As everyone was watching me, including Angie, no one had noticed him climb into his truck and start backing onto the road.

I took a few deep breaths and continued my descent, but I couldn't see anyone in or around the car. I was mystified. The only sign of life was a child's blue shoe lying in the dirt far below.

Then a voice called, "We're over here!" Off to the left I could see a group of four people standing on a ledge under a large rock overhang—they appeared to be a mother, father, young girl and even younger boy. From above they had been completely invisible. The boy pointed. "Can you get my shoe?"

"Is anyone hurt?" I asked.

The father replied, "No, thank God! We have a few bruises—that's all."

We spent the next hour and a half lifting the family up the cliff one at a time in the basket stretcher. The shoe is still there.

The family was visiting from India. I don't know what their religion was, but I'm willing to bet that if they didn't believe in miracles before the accident, they do now.

HANGMAN'S FRACTURE

While I was working in that little town, I helped organize a number of fundraising drives to buy rescue equipment. The community was very supportive, and we were able to purchase all sorts of things, right up to a rescue truck. Of course, as soon as we purchased our new truck and organized a search-and-rescue group, the semitrailers stopped crushing cars on the highway, and for over a year the truck didn't turn a wheel.

However, one piece of equipment the community helped us acquire was really handy. Although the Kendrick Extrication Device, or KED, is rarely used on most city ambulances, it is very useful in car accidents because it can be fitted around the neck and chest of the victim while they are still in their vehicle helping to brace their spine while they are being extricated.

As soon as we received our KED, we held a training class and then put it in the back of our main ambulance. The very next day our first call was for a single-car accident twenty kilometres up the highway. At the scene we found a very anxious woman holding her neck.

"It doesn't feel right," she said. "I think it's really hurt."

We immediately trussed her up in our new gizmo, paying particular attention to her neck, and ran her north to the closest hospital. There we waited around while the doctor had her X-rayed right through the KED.

When the film was developed, the doctor let out a whistle and called us over. "Look at this break at the base of the skull," he said. "A hangman's fracture! Executioners put the knot of the noose right there to kill their victims instantly. Almost no one ever survives this because any movement of the head will sever the spinal cord. You did good work bringing her here in one piece."

"It's more than good work and more than good luck," I replied. "If she'd had her accident just one day earlier, we wouldn't have had the proper equipment and would probably have killed her getting her out of the car. She's only alive because the angels were helping."

SPEED BUMP

Some accidents aren't caused by bad luck but by stupidity.

We were sent out around ten at night for an eighteen-wheeler that had missed a turn and run off the highway. By the time we arrived, four or five pickup trucks were already at the scene. Most of the drivers were busy scavenging crates of wine from the big trailer, which had jumped the ditch, flipped on its side and split open. Fortunately, one man was more responsible, and he guided us with his flashlight farther into the bush to where two big fir trees had brought the cab to a crushing stop.

The driver was inside, his feet pinned under the pedals. He was conscious and groaning in pain from multiple fractures to his legs and left arm. After bracing his neck, we gave him Entonox for the pain, carefully bent back the pedals and took him out of the cab on a backboard. As I was splinting his fractures, I said, "Don't worry. We'll have you out of here in a minute and on the way to the hospital. They'll get you back on your feet."

"No, wait!" the trucker said. "I have some pills in the glove compartment. Can you get them for me?"

"What kind of pills?" I asked. "You said you weren't taking any medications."

"They're bennies [Benzedrine, a form of amphetamine]. I've been driving non-stop for thirty-six hours so I was taking them to stay awake—but I guess I nodded off. Can you get me the pills? If the cops find out I was doing speed, I won't be able to collect insurance, and I'll lose my licence."

I climbed into the cab, found a small pill bottle and gave it to him.

He put it in his shirt pocket. "You won't tell the cops, will you? I have a family to feed, and if I lose my licence, I won't be able to work."

I'm a sucker for a sob story. "Okay," I said. "It's none of my business. I expect you've learned your lesson."

We carried him out to the road and loaded him into the ambulance. By now the police had arrived and the pickups were scattering. A cop asked me for the trucker's name before saying, "I'll let you get going to the hospital. I'll follow once I've evaluated and secured the accident scene. I can ask him questions after they've treated his injuries."

The cop closed the back door, the trucker gave a pained sigh of relief and we drove off with our patient and the evidence.

Now I'm sorry I didn't turn him in. After all, he could have easily killed someone when he fell asleep. Is he still driving long hauls and still taking bennies to help him get through the nights? I don't know.

KARMA

There was a bit of an incline where the van had gone over the cliff, so it had bounced on the way down, ending up ten metres below, lodged against a rock. The woman driver was lucky to be alive, but the fall had fractured her spine. She was not a happy woman as we strapped her to a backboard, lifted her up the cliff in a basket stretcher and put her in the ambulance.

That was when she started to cry. "This is really bad, really terrible!" she said through her sobs.

"No," I said. "I'm sure it's painful, but it's not terrible. You don't have any loss of sensation, so you probably don't have any major nerve damage. You should make a full recovery."

"No," she moaned, "you don't understand. It's all my fault. My van was breaking down and I don't have the money to fix it. So I decided to drive it off the cliff and collect the insurance, but I didn't get out in time. And now I'm going to get charged with fraud."

She was right. She really had made a complete mess of things. She had hoped the van would drive slowly over the cliff by itself, but as soon as she jumped out, the engine stalled. So she tied a string around a big rock and put it on the accelerator. But when she was about to jump out the second time, the rock got stuck between the pedals, and while she was pulling the string to get it loose, the van went over the cliff. Now she had a broken back and a dodgy-looking rock was lodged inside the wrecked van.

The Mounties don't always get their man or woman, but this time they did. She was charged with insurance fraud and spent two months in jail. She lost her job, and it was another eighteen months before she could walk properly.

Talk about dumb! There's no point breaking your back for a few thousand dollars. Petty crime is a mug's game. If you are into fraud, you should get a job on Wall Street and learn from the experts. Then you can steal billions, and if you bust the bank, the government will bail you out. Win or lose, you will never go to jail.

Saved by the Yell

As a small-town unit chief, I was often on call twenty-four hours a day, seven days a week. One evening I finally got to bed just before midnight, exhausted after a long workday followed by hours of pacing the floor with a colicky baby. Half an hour later my pager went off. It was a routine call for a sick woman, but she lived on a farm far off the main road, and it took my partner Angie and me almost three hours to pick her up and transport her to the city hospital. I then drove the ambulance to an all-night service station, filled up the tank and started the long drive home.

It was Angie's yell that snapped me awake. I had been driving the ambulance in a straight line, but the road had turned to cross a bridge, and we were just about to plunge down a steep bank into the river. I wrenched the steering wheel to the left and swerved back onto the road, just missing the corner of the bridge. Once across it, I pulled over to the shoulder and stopped, my hands shaking. I turned to my partner. "I'm sorry. That was really close!"

Angie opened her door and walked around to my side. "Get out," she said. "You're fired! From now on I'm driving."

Thanks, Angie. You saved our lives.

WHISTLING IN THE BODIES

It was an odd but tragic accident. On a sunny day two cars collided head-on on a straight stretch of highway. In one car was a couple from Kelowna; in the other were four Australian tourists—two retired couples who were making their first visit to British Columbia. We would never know who was at fault because the BC couple had life-threatening head injuries and couldn't remember what happened. The two Australian men were sitting in the front of their car and were killed on impact. Their wives survived with moderate injuries because they were in the back, but they had no idea what had happened because they were chatting at the time.

As soon as my partner and I received the call for a two-car head-on collision, I asked the dispatcher to page our town's second ambulance as well. He told us that the police were already on their way and he would be backing us up with an ALS unit from the city.

The cars were badly buckled, and it took us thirty minutes to extricate the victims. I sent the critically injured couple off in the ALS car and in our main ambulance with one of the cops as the driver. We then loaded the Australian women into our second car. "Are our husbands all right?" one of them asked.

I lied to reduce the shock. "I'm afraid they've also been injured—I don't know how badly. We've sent them ahead to the hospital."

In reality, the bodies of the dead men were lying by the side of the road, covered with brown blankets. Twenty minutes earlier I had given dispatch a scene assessment and been told that another vehicle was being sent to pick up the bodies. So now I stood waiting in the sunshine, watching a Mountie wave traffic past the wrecked vehicles. The disciplined rush of the call was over and had been replaced with sadness as I looked at the bodies and thought of the heartbreak ahead for their families.

It took another half hour for an old ambulance to appear, and to my surprise it was driven by Joe, our cantankerous dispatch supervisor. "I can't believe I'm doing this," he said. "Almost every car in the region is busy, so I had to dig out this antique and drive it myself."

We put the dead men in the back. Joe turned the ambulance around and we headed back to the city. He wasn't very cheerful, and after a few minutes he said, "This business made me miss my lunch, and now I really need some food, some coffee and a cigarette. I'm not in the mood for a long drive so I'm going to speed this trip up." He reached on top of the old-fashioned dashboard, found the switch controlling the emergency lights and turned them on. Then he stepped on the accelerator and we raced ahead, with other vehicles obligingly pulling out of our way.

It didn't take long to reach the city. As he turned onto the circular road that led to the hospital's emergency department, I said, "Joe, you still have your lights on. You'd better turn them off."

"Oh, right," he said and reached for the switch. But instead of turning off the emergency lights, he hit the siren, and it started to wail just as we turned the corner in front of the emergency department. Flustered, he tried to turn it off, this time only succeeding in turning on the fog lights. He hit more switches and eventually the wailing ceased. He pulled to a stop beside the main entrance—and turned red from embarrassment.

There on the sidewalk stood the regional manager. As this was a relatively serious incident, he had come to the hospital to make sure everything went well, and now he was waiting for me to arrive and give my report.

He strode up angrily and yanked open the driver's door. "Joe, have you lost your mind? Why are you driving with the siren on in a hospital zone? And what the hell are you doing whistling in dead bodies? Have you gone crazy on me?"

Joe didn't have an answer. Completely lost for words, he could only wave his hands and sputter.

The situation was actually hilarious, and although it is never a good career move to laugh at your superiors, I almost cracked up. But then I thought of those poor wives lying in the emergency room, who would soon learn the truth about their husbands. What a terrible way to end a holiday.

THE UNLUCKY TRUCKER

Whenever I worked with Dan, we inevitably got into swapping stories, so one day while I drove the ambulance toward a hospital pickup, he told me the tale of an unfortunate patient in Dawson Creek.

It seemed the local part-time crew had been paged to respond to a single-vehicle accident some thirty-two kilometres out of town. The first paramedic raced down to the station, drove the ambulance out of the bay and waited for his partner. Three minutes went by, then five. The driver phoned dispatch and had his partner paged again. Another five minutes passed, and then ten.

At this point the frantic driver phoned his partner's house, only to discover that, unaware of the new work schedule, the man had checked out of town on a fishing trip. The driver tried the numbers of a couple of other part-timers, only to draw blanks.

What should he do? The dispatcher was getting pretty antsy about the delay. After another minute of confused indecision, he jumped back into the driver's seat and tore out of town alone.

At the accident scene he found a semi that had skidded off the gravel shoulder and careened into a rock face. Beside the tilted cab a big trucker sat nursing a broken foot. He didn't look very happy. "That took you long enough!" he complained. "You drive all the way from Prince George or something?"

"Don't worry," the paramedic said. "You'll be at the hospital in no time." He cut the trucker's sock off, wrapped the broken ankle in a pillow splint and then looked around for help from a passerby. Nobody! The trucker had called in his own accident on his radio. "Sorry, buddy," the paramedic said, "but I'm on my own here today. You're going to have to hop over to the ambulance."

"What the hell? Can't the government afford two medics? Oh, all right, lemme hang onto you then."

Cursing and groaning, the trucker hopped up out of the ditch and into the back of the ambulance, finally lying on the stretcher with a sigh. "Okay," he grumbled, "let's get to a hospital quick. My foot feels like it's about to fall off."

The paramedic had just turned the ambulance around and started back toward Dawson Creek when dispatch called. "There's another MVA twenty-six kilometres farther out. You'll have to turn around and go back and check it out."

"Ten-four," muttered the paramedic. Then raising his voice, he told his patient, "Sorry, buddy, but we have to check out another accident. We'll be just a few minutes longer."

"Oh, great," came the reply. "They'd better be dead already 'cause I'm getting in the mood to kill somebody."

The car was flipped over in a ditch alongside a lonely stretch of road. A passerby had spotted the accident and driven off to find a phone, leaving an unconscious man inside the car, his face and head badly lacerated. The paramedic stabilized his injuries, dragged him out of the car onto a backboard and strapped him down. Then, sweating and straining, he hauled the board with his trussed-up victim out of the ditch, across the gravel and up behind the ambulance.

"Sorry again, buddy," the paramedic said, "but you're going to have to give up your nice bed for this guy. Uh-h ... and I hate to ask you, but do you mind hopping down here and giving me a hand? This guy's too big for me to pick up by myself."

Swearing and grimacing, the big trucker hopped down and helped manhandle the unconscious man up into the back of the ambulance. "Now what?" he said. "You're not leaving me back here with this guy? He looks like he's ready to croak, and I don't need a murder charge along with a broken ankle."

"Don't worry—I'll be taking care of him." The paramedic tried to look calm as he wiped the dirt and blood from his pants. "You're going to have to drive."

"Drive? With this big pillow on my foot? I can't even fit behind the steering wheel."

"Well then, just take the damn thing off!" The paramedic was losing his sense of humour. "Don't be such a wimp. Just get going."

So the trucker got to drive his first emergency call back to the hospital in Dawson Creek.

Dan and I laughed when we thought of the expression the poor man must have had on his face when he received the bill for his ambulance trip. The government probably charged him double for all the extra kilometres he had to drive!

Quick Fixes

There are some people you never forget.

Jack was the ambulance unit chief in the next town up the highway. He was a keen athlete and built like an ox. When he was younger, he had won a scholarship to play football for a top American university, but shortly after he crossed the border, his car ran into a truck. A fractured leg ended his dreams: the scholarship was withdrawn and he went back home. After he recovered, he had to pick a new career. He became a paramedic.

One sunny afternoon both our ambulances were sent Code 3 to a two-car accident that had occurred halfway between our towns. We arrived at the same time. The first car was by the side of the road, its occupants shaken though unhurt. But the second car had jumped the ditch and hit a rock. The driver's face was covered with blood, and he appeared to be unconscious.

Jack and I tried to open the doors, but the car frame was bent and the roof partially caved in. "We'll have to cut him out," I said. "I'll tell Brad to get the Jaws from our car."

But Jack said, "Don't bother." He reached down, grabbed the twisted driver's door and pulled. The metal door groaned and then tore free. Jack grunted and threw it behind him. "All right," he said. "Now all we need is a backboard."

I was astounded. Had I really seen Jack rip a car door off its hinges? But it had happened. Brad had seen it too.

Too bad about that football scholarship. He would have been unstoppable. But even without football, Jack was a pretty happy guy. He loved his family, his job and his community. The only thing he didn't like was his office. It was a claustrophobic, windowless hole on the second floor of the ambulance station. He longed for light, fresh air and a view of the beautiful mountains.

So he sent a memo to Hugh, the regional manager, requesting permission to install a window in his office. (In those days inter-office correspondence was all conducted with memos—in triplicate.) But Hugh refused, explaining that although money was available for repairs, there was no budget for renovations. On the bright side, however, he pointed out that Jack's town was scheduled to get a new ambulance station in eight years.

Jack had no intention of waiting that long. One day the following week he came to work with a sledgehammer and knocked a huge hole in the wall. Then he sent another memo: "I regret to inform you that I have a large hole in the wall of my office and the rain is getting in. Can I have it repaired?"

A few days later he received a reply: "Fix the wall before the rain causes further damage. Get three estimates, have the hole repaired and send us the invoice." Jack got on the phone right away. To no one's surprise, it turned out that the cheapest way to fix the hole was to install a window.

Unfortunately, Jack didn't get to retire—he died from cancer. A good and unforgettable man.

The Friend I Avoid

The girl was bicycling home in the evening when a drunk driver came tearing out of the bar's parking lot, rounded the corner on two wheels and knocked her twenty metres across the road and over a bush. By the time we got there a cop was already doing CPR.

"I don't think she'll make it" were his first words. "She's been hit really hard."

I felt her skull, and the back of her head was mush. Although I knew there was probably no hope, I decided to continue cardiopulmonary resuscitation. We scooped her into the ambulance, contacted the town doctor and headed to the medical clinic a few blocks down the road. There we waited for the doctor to arrive.

I was doing one-person CPR in the back of the lit ambulance when the girl's mother ran up. Looking in the rear windows, she saw me working on her daughter and immediately realized what had happened. I will always remember her howl of anguish, and I knew she would never be able forget the moment when she learned that her daughter was dead. The doctor arrived and called off efforts to revive the girl. She was buried, the driver was charged and the mother went through a period of intense grieving. After a few weeks she went back to her job at the local supermarket (where I shopped) in order to support herself and her remaining daughter.

As for me, my life continued as usual, except that I could no longer walk comfortably around the town. I had been friends with the mother before the accident, but now whenever I happened to see her, a look of shock would come over her face. Much as she would try to hide it and be pleasant, I knew that each time she saw me she was reliving the moment when she saw me, my hands covered with blood, working on her daughter in the back of the ambulance.

One of the reasons I was glad to transfer back to the coast six months later was that I would no longer be a constant reminder of her loss and pain.

Breakup

At the time, I didn't see the collapse of my marriage coming, but in retrospect I can see that it was almost inevitable.

I liked living in a small town. I had not only succeeded in building a better Ambulance Service there, but I had organized a community development group. We had ambitious plans to line the main street with traditional wooden façades and boardwalks and make our district a western-themed, four-seasons holiday destination. Some of the best artists in Western Canada lived in the area, and two of them had agreed to paint murals on the sides of downtown stores to complement the design.

Our proposal called for paths leading to the centre of town that could be used for horse riding in the summer and cross-country skiing and snowmobiling in the winter. Although most of the local businesspeople were enthusiastic, the restaurant owners were outraged about our plan to put hitching posts outside their front doors: "And who's going to clean up all the horse poo? Not us!" I told them that if they let us spruce up the town, every tour bus would stop by, their business would double, and one of the extra employees they hired could clean up any droppings, but they wouldn't budge. "You call this a business plan? All you're going to attract is horseshit and flies. No way!"

Despite the occasional problem, I was happy, but my wife wasn't. Karin had grown up in a big European city, and although she made friends in our adopted town, her family and closest friends were far away. Also, I admittedly wasn't very romantic. We had married during my paramedic training, and while I had skipped an autopsy to attend the ceremony, I had turned down an offer of three paid days off because I didn't think I could afford to miss more classes. So we never had a honeymoon. And after we moved to the Interior, I was always on call so we could rarely get away, and when we did, we were never alone because we were looking after small children. I wasn't worried because

I thought it was normal for young parents to be always busy and tired, and I figured we should and could just tough it out.

But Karin and I gradually drifted apart. I should have signed us up for couples counselling, taken time off work, gone away with her on an extended holiday. Instead, sure that everything would improve with time, I soldiered on, until one day Karin told me that she was taking the kids and moving to Victoria.

What a depressing mess. I'd lost my family and, with them gone, my reason for living in the Interior. To cut expenses, I sold our house and moved into a little hut in the bush, hoping that soon I would be able to transfer to a unit chief position in the Victoria area. But summer turned to fall and fall to winter without any unit chief vacancies being posted.

Part IV: Victoria

It was the end of March before a position became available in Victoria, and although it was only for an intermediate-level paramedic, I took it, loaded my possessions into a trailer and headed for Vancouver Island. Because I was taking a wage cut, I rented a house with three other single men and put bunk beds in a corner of my room so my kids could stay with me on my days off. It was time to start over.

The Right Credentials

It wasn't my first attempt to move back to Victoria; a couple of years earlier I had flown there for a promotion interview. At that time my parents lived in Victoria, and since I didn't have a ride from the airport, my father picked me up at the terminal and drove me to my appointment. He was proud of his British background and always drove a white Jaguar, which nicely matched his white hair and white moustache. So I travelled in comfort into town, where he let me off in front of the regional office.

I walked inside to see one of the supervisors staring out the window.

"That's the Lieutenant Governor!" he exclaimed.

"What? Who?" Confused, I walked over to the window to see who he was looking at, just in time to see my father driving away.

"That's Lieutenant Governor Rogers!"[1] The supervisor looked awed. "How do *you* know *him*?"

---

1 Robert Gordon Rogers was lieutenant governor of BC from 1983 to 1988.

At that point I blew it. I could have said, "Oh, I saved his life a while back," or perhaps, "He liked an article I wrote on improving health care in British Columbia, and we've been corresponding ever since." But instead, off guard and honest, I replied, "No, you're mistaken. That's my father."

"Just your father?" The supervisor was embarrassed and miffed. "But he does look a lot like the Lieutenant Governor." And he dismissed me to a chair in the hall. The interview went downhill from there.

I've often wondered what would have happened if I'd complimented that supervisor on his photographic memory instead of correcting him. I'm sure word of my direct connections to the top would have spread around the office in two minutes and my chances of being promoted greatly improved. It might have changed my whole life—more status, a bigger salary, a nine-to-five job in a stuffy office, endless meetings, less hair and the beginnings of a paunch. A very different life from front-line work in the community.

Well, I didn't feed him a story and I wasn't promoted. For the rest of my career as a paramedic I remained your basic shift-working paramedic charging around town like a loose cannon. I never had to worry about office politics and life was never boring. Perhaps honesty does pay.

Management

We rarely saw any Ambulance Service managers. Buried in paperwork, they left most day-to-day administrative responsibilities to the frontline unit chiefs. As long as we did our jobs and they didn't receive any complaints, they left us alone. Which was just fine with me.

We knew why they were always busy. After dropping off a patient in emerg, we were putting fresh linen on the stretcher when Mark said, "Make sure you smooth out all the wrinkles. Management is worried about the bottom line!"

Of course they were worried; like most government health services, the Ambulance Service was always over budget. To try to get a grip on expenses, at one point the provincial government sent three accountants to the BCAS head office to find ways to reduce Advanced Life Support overtime costs. A senior manager asked them, "What would you like us to do, shut down the ALS cars after midnight on weekdays or on weekends?" Two of the bean counters pulled out their calculators to work out the most cost-saving option, while the third leaned back in her chair and laughed. You can't shut down emergency services—the public wouldn't stand for it. Whenever they run out of money, the politicians just have to find a little more.

The management of the British Columbia Ambulance Service can take credit for building one of the world's biggest and best ambulance services in a remarkably short time. Nevertheless, from time to time problems have arisen.

PIP, PIP AND ALL THAT

Every now and then BC's emergency services hold joint meetings to co-ordinate disaster planning. In the early days of the BCAS, some ambulance managers felt they were not taken seriously at these gatherings because senior police and fire personnel looked down on them as amateurs belonging to a junior, Johnny-come-lately service. Part of the problem, according to the ambulance station chiefs, was that they didn't *look* equal to the other services. Their uniforms had just two silver stripes on each epaulette, while fire and police chiefs had three gold stripes to indicate their rank, and their senior managers sported shoulder pips like military officers. So the ambulance chiefs kicked up a fuss until they were issued three gold stripes too.

But this created new problems. If front-line management had three gold stripes, then upper management would have to be issued with something better—at least four gold stripes or pips. Soon supervisors started to wear increasingly imposing uniforms as well. However, some were not happy being outranked by their colleagues, and a war of gold braid broke out. This competition reached its logical conclusion when one day the public relations manager appeared on the evening news decked out like a third-world dictator. Fortunately, calmer heads prevailed and after a few weeks, ambulance managers went back to looking like emergency personnel rather than members of an invading army.

OUR MITEY MANAGERS

Every organization, whether private or public, develops its own culture. The BC Ambulance Service is no different. Most of the time its many components work so smoothly that you can almost forget that they exist: vehicles are maintained, new staff are hired and trained, wages and bills are paid, cars are crewed, emergency calls are answered, patients are treated and transported—not only 24/7 but on time and to high standards. Both the public and its employees are looked after with care and kindness.

However, although the Ambulance Service and its staff are usually efficient and occasionally brilliant (in crises, for example), they aren't perfect. In my days as a paramedic some things barely functioned. For example, over time a regular procedure had developed for employee–management dialogues. First, you requested a meeting with a manager at the regional office. There you presented what you believed to be the facts, the paperwork went back and forth, they denied your case, and then you went to the union and submitted a grievance. The process could also be described like this: you started off friendly and trusting, next you became confused by the response, then you got defensive and finally totally pissed off.

My own issue concerned having contracted scabies twice in the same year from transporting patients from the same extended-care hospital. These microscopic mites can easily burrow into elderly people's dry, cracked skin, and once an infestation spreads through a care facility, it is very difficult to eliminate. Although the nurses denied having a problem, I didn't believe them because everyone in that hospital—staff included—was constantly scratching.

As paramedics, in order to lift patients out of their beds and onto our stretchers, we had to hold them next to our own bodies, so despite wearing protective gloves, the bugs managed to climb on and crawl into my skin. What misery—I was too itchy to sleep! And each time it happened, the process of ridding my body, clothes and bedding of the tenacious mites took weeks and cost me $250 in ointments and dry cleaning. The first time I covered the cost myself, but the second time I decided it would only be fair if the Ambulance Service picked up the tab, and I asked to meet with a supervisor at the regional office. As usual, it didn't take long before he was making me feel like an idiot for not having adequate documentation. "Can you prove that the mites came from that particular nursing home?" he asked.

"I don't know," I replied, thoroughly frustrated. "I guess I'll have to compare the DNA of their mites with the dead specimens lying around my house."

I thought he was just being absurdly bureaucratic until he made a comment that pushed me over the edge. Leaning back in his chair, he said, "You say that you had scabies and therefore had to buy non-prescription drugs, but why should we trust you? You didn't see a doctor. How do we know you aren't just billing us for drugs that you're going to peddle on the side?"

Now, I've done a lot of travelling and had many people try to sell me stuff on the street. In San Francisco hippies would walk by muttering, "Acid, peyote, mescaline, anything you want, cheap!" In Morocco twelve-year-olds would offer hash and marijuana in ten languages: "Come with us to the top of the old fort and watch the pretty sunset. You can try a free joint!" In Toronto and New York strangers leaning against walls have opened boxes stuffed with smorgasbords of drugs. But nowhere has anyone ever sidled up to me and tried to sell me scabies ointment. I tried to imagine what was going through the mind of the supervisor: ambulance crews walking along Government Street in Victoria and whispering to the tourists and pimps, "Pssst, you want a hot tube of scabies cream? Only $5!" And you can imagine the customer response: "What kind of shit are you talking about, man? This some kind of crack? You smoke this stuff or what?"

Those little bugs sure got under my skin, but what had happened to that supervisor to make him so paranoid?

The Art of Ambulance Driving

Any fool can drive, but it takes a certain amount of skill and luck never to hit anything. Although I never considered myself to be a brilliant paramedic, I always took pride in my ambulance driving skills, and after nine and a half years on the job my record was unblemished. I had driven at recklessly high speeds through traffic jams and snowdrifts without once scratching the paint—let alone killing anyone. So I was looking forward to receiving a public award for surviving ten years of breaking every traffic law ever written.

But fate cruelly intervened. I was pulling away from a curb when I heard an odd screeching sound. A fire hydrant had been cunningly positioned so that it hung over the gutter, and it had worked like a can opener, ripping open the back end of the ambulance and shredding my one claim to professional competence. I was disheartened, but not enough to change my driving style, and I continued racing through narrow gaps between cars with only a couple of centimetres to spare on either side. There's a fair bit of pleasure in knowing that the occupants of the cars you are aiming at are watching you in horror as you bear down on them, only to be amazed to have you slide past them and disappear on down the road. I can imagine the wife turning to her husband to say, "Wow, Stan, now that's good driving. You could never do that—it always takes you three tries to parallel park."

"You call that good driving?" he says. "That guy should be locked up!"

But despite my faultless slaloming around city streets, my reputation never recovered after that minor mishap with the fire hydrant. I took to arguing with the garage doors of ambulance stations. (Click the automatic door opener once and nothing happens. Click twice, nothing. Click three times and the door goes up, only to remember your previous instruction and come down on top of you as you back in, etc.) My colleagues began to post warnings about me, and my self-esteem took a beating.

Province of British Columbia

MEMORANDUM

TO: Graeme

FROM: John + Ken

SUBJECT: Vehicle Operations

DATE: 16 Jan. 93

FILE: Crunch

☐ For Your Information ☑ Please Process ☐ Please O.K. and Return ☐ Return With More Details ☐ Please Discuss With Me ☐ Investigate and Report ☐ Per Your Request ☐ Please Answer ☐ For Your Signature ☐ For Your File

Due to the unblemished nature of the exterior of this brand new car, would you kindly refrain from driving into Ambulance stations, bay doors, elephants etc!

Thank you for your anticipated co-operation

Love "C" Platoon Bravo/Charlie

REPLY:

WRITE YOUR REPLY AND RETURN THIS SHEET

However, after fifteen years I still hadn't hit anyone or any other vehicle, and I began to convince myself that I wasn't a total klutz. That is, until nine o'clock one black night.

BUMPER CARS

I was driving with my partner, Mark, beside me on a routine call for an unknown problem. It was our last night shift and we were both tired and snarly, so we snarled ourselves deep into an argument. (Sometimes I think we really belonged in different worlds: he would prefer to live in the simpler age of steam engines, whereas I wish I was living in a more progressive era.) This time we were arguing about feminism, and as I remember Mark was being particularly stupid, though I'm sure he recalls the matter differently.

Because my attention wasn't on the call, I drove right past the patient's house, not realizing my mistake until I was halfway down the block. Swearing, I put on the brakes and asked Mark to shine the spotlight on the house numbers on my side. Instead of holding the light outside his window and shining it over the cab, he shone it across my face and out my window, half-blinding me.

"Back up three houses," he said.

Completely pissed off with him by now, and with spots in my eyes, I glanced in both my mirrors to see whether anything was behind. Nothing but blackness and more spots. I put the transmission in reverse and stepped on the gas. *Wham!* The ambulance hit hard. "Oh, God!" I groaned. "There goes my licence. Mark, take the jump kit and check on the patient. I'll see what damage I've done."

I got out and walked back to where an attractive woman was standing next to an expensive new car. It was tucked in right behind the ambulance, where I couldn't see it in the mirrors. But I had no excuse. I should have seen the reflection of her lights, and more than that, I should have asked my partner to get out and direct me while I backed up. In the Ambulance Service, you can drive almost any way you like, but if you cause an accident, they let you hang out to dry.

"I'm terribly sorry," I said. "It's completely my fault. I didn't see you."

"No, no," said the woman. "It's my fault. I should never have driven so close to an emergency vehicle."

"No," I repeated. "I'm entirely to blame. I should have checked before I backed up."

"No, no," the woman protested. "I should have kept much farther back from you."

We were getting nowhere, and I was on a call. I hoped our patient wasn't bleeding out in Mark's arms. "I'll get a flashlight." I rifled under the driver's seat and brought one back, afraid of what I'd find.

To my amazement everything looked normal. Not a broken light or a ding on either vehicle. I looked underneath and on either side. The bumpers must have absorbed the heavy shock without even marking the rubber. "It looks fine," I said, "but let me give you my name and registration for the police report."

"No, that's all right," said this surprising lady. "I don't want to keep you from your work any longer. I'm sure everything is all right. I'm sorry I caused all this trouble." And after a few more mutual apologies, she drove off.

I backed up carefully and went in to help my partner. It was a minor problem—an arthritic older woman had slipped to the floor and couldn't get up. We put her back in her chair, wished her a pleasant evening and left.

Back in the car, I was still shaken. "No damage was done, so I guess there's no need for any paperwork," I told Mark. "Perhaps ignoring the whole business is the easiest solution."

"I don't know," he said. "You can get in a heap of trouble if you don't report an accident in a government vehicle. But you're the driver, so it's up to you."

I debated the issue with myself, but in the end I decided no one would notice if didn't write it up, and I just went home at the end of the shift, relieved to have four days off.

I came back to work to find a note on the steering wheel: "Would whoever bent the back bumper report the accident and fill out all the incident forms." Oh-oh. Now the shit was going to fly. I had failed to report an accident and been caught.

Nervously I went back and looked at the bumper in the daylight. It had no dents, but now I could see that the whole bumper was bent down a few centimetres. No need to panic, I told myself. I can scam this one easily. I borrowed a large jack and levered the bumper back into alignment. There, no more accident! At the end of the work block I left a note for the incoming crew: "No problem, no paperwork." And I went home, the matter forgotten.

Back at work the next week, I was shocked to find a brown letter from the Ministry of Transport waiting for me in my office mail slot. The game was up. I had failed to report an accident, and even when caught, I had refused a direct order to fill out the paperwork and instead tried to hide the damage. The least I could expect was to have my driver's licence suspended. But in all likelihood I would face disciplinary action as well as the possibility of my entire career as a paramedic going down the tubes. With trembling hands I opened the envelope. The letter began: "In recognition of your many years of safe professional driving, the Government of British Columbia awards you this Safe Driving Certificate ..."

"Yes! I've done it!" I told myself. "There is an art to being an ambulance driver, and I've mastered it at last!"

My Quickest Call

We were heading back to the station, Mark driving the ambulance and me sitting in the passenger's seat. Directly ahead of us a man in a bright yellow helmet was squeezing his bicycle between the traffic and the vehicles parked beside the curb.

Suddenly the driver of one of the parked cars opened his door. The cyclist crashed into it, flipped up in the air and landed on his back on the pavement.

Mark pulled up beside him. I opened my door, leaned out and looked down. "Are you okay?"

The cyclist opened his eyes and stared at the ambulance in amazement. "Damn, you guys are fast!"

MAKING A SPLASH

While most people are aware that ambulance driving is difficult and dangerous work, few understand the co-ordination needed to drive a Code 3 through city traffic at five on a Friday afternoon, weaving in and out among the crowded vehicles while simultaneously keeping your cup of coffee from spilling as you answer the radio. When you can do that without wrapping the microphone cord around the steering column, you've graduated as an ambulance driver.

The job is made more difficult by the fact that there are no scheduled breaks for coffee or meals. The good part is that you get paid straight through, and if the day is slow you get to drink a lot of coffee. The bad part is that if things get busy, you may work six or eight hours with no breaks at all, and that gets pretty tiring.

In busy stations, you always bring a packed lunch so you can grab a bite to eat between dropping one patient off at a hospital and going to pick up another. You also grab drinks whenever you can, often having to either dump them out at the side of the road or slurp them down quickly when an emergency call comes in. I once picked up a couple of milkshakes and set them on the engine cover only to have a call come in and my partner pull a sharp left. We were swimming in pink for the next hour, but at least the shake missed my uniform. One of the other paramedics, though, had just started on an extra-large cup of coffee when the driver accelerated sharply. He attended that call with a large brown stain dripping from his crotch.

NEITHER SNOW NOR RAIN ...

When a call comes in, you are expected to get there—somehow. It doesn't matter whether the patient is stuck in a tree, has fallen into a canyon, or is at the end of a dirt trail a mountain goat couldn't traverse—the paramedic is supposed to arrive promptly with a full load of equipment. People call us when half a metre of snow has fallen and they don't want to risk driving anywhere, or when a blizzard is blowing and the visibility is zero. In this country somebody ought to supply us with snowmobiles for winter travel instead of rear-wheel drive trucks, but they haven't, so we do what we can.

A paramedic from the Kootenays told me of going to an accident that had been caused by black ice. A car was wrapped around a tree at the bottom of a long, steep hill, and as soon as the ambulance nosed over the crest of the hill, it slid sideways into the ditch. It was impossible to stand on the glassy road, so the attendant got out, put her jump kit on the ground, sat on it and slid all the way to the bottom. But it took a tow truck to get her, the patient and the ambulance up to the top of the hill again.

SLIP-SLIDING AWAY

Sleet had been falling all afternoon, and with vehicles piling up all over Victoria, my partner Mark was driving a Code 3 along Blanshard toward an accident downtown. Three lanes of cars were stopped in front of us at the Hillside intersection, so Mark gently touched the brakes. No use! We might as well have been driving across a hockey rink. The ambulance just kept sliding toward the rows of stationary cars.

"Mark, jump the meridian!" I pointed left.

He turned the wheel and the ambulance bounced up onto the divider, back down on the other side then continued to slide right across the three opposing lanes until it bounced up onto the sidewalk. It came to a stop beside the wall of the Memorial Arena. With his siren still wailing, Mark backed up a few metres, pulled back out on the road, cut across the waiting traffic and continued south down Blanshard. I can imagine how the other motorists must have cringed as they sat watching our big truck sliding sideways toward them, but we managed to look calm and professional while we skated around. After all, we have reputations to uphold—anyone who drives an ambulance in winter has to stay cool.

SMOOTH DRIVING

One beautiful summer afternoon I was washing our ambulance outside our station while in the next bay Dave washed the Advanced Life Support ambulance. He had placed a mug of coffee on his vehicle's left siren, and he would take a sip of coffee, then dip his long-handled brush into a bucket of soapy water and scrub for a while before pausing to drink some more coffee.

The hotline rang and then stopped as it was picked up inside the station. Thirty seconds later Dave's partner opened the door. "It's for us, Dave," he said. "Code 3 for chest pain." They jumped into their ambulance and pulled away, lights flashing.

"Wait!" I yelled. "You left your mug on the siren!" They didn't hear me. Their car picked up speed and disappeared around a corner, and the wailing of their sirens faded away. Five minutes later they returned and parked in the bay again. Dave opened the driver's door and stepped out.

"We were cancelled," he said.

Curious, I looked at his siren. To my amazement the coffee cup was still sitting on top, right where he left it. Dave had driven a fast Code 3 without it falling off. Now that's smooth driving!

THE PENDER ISLAND TREASURE MAP

Mark and I had been asked to drive a new ambulance to the Swartz Bay Ferry Terminal, where we were to exchange ambulances with a crew of part-timers from Pender Island. They would take the new ambulance back to their island while we would drive their old vehicle to a depot in Victoria.

We were about halfway back to Victoria when dispatch called us. "I have a Code 3 for a three-car MVA, and you're the only free ambulance in the area. Are you able to do the call with that wreck? Has it been stripped bare or do you still have a little medical equipment?"

"We'll manage," I said. "We have our jump kit and a stretcher."

"Good," said the dispatcher. "It's a Code 3 in the 3000 block on Lochside Drive."

"Ten-four," I replied. I turned to Mark. "Isn't 3000 Lochside in Cordova Bay?"

"It should be," Mark said, "but I'll check the map." He started looking around the cab for a city map. In the meantime, I flipped on the lights and sirens and accelerated down the Pat Bay Highway toward Victoria.

"They've emptied the car," Mark said. "There's not a map in sight. Oh, wait, I spoke too soon—they've left us something." He took a folded piece of paper out of the glove compartment and opened it up. "It's a treasure map of Pender Island!"

I laughed. "That's a big help. Never mind; we know where we're going." But a few minutes later I said, "You know, just in case, I'd better ask dispatch for a cross street."

The dispatcher was aghast. "Cordova Bay? You're going to Lochside Drive in Cordova Bay? The accident is in the 3000 block of Lochside Drive in *Sydney*! You had better put the pedal to the metal—it's a three-car MVA and you've been going the wrong way."

Good Lord, we had just lost ten minutes! "I hope no one is dying," I said to Mark. I crossed the meridian and pressed the accelerator to the floor. In response the heavy old ambulance leaped forward, roaring and rocking. Breaking into a sweat, I gripped the wheel tightly as we raced back up the highway. "The traffic had better get out of our way—I won't be able to swerve without flipping this thing over."

Mark looked dubiously at the speedometer and tightened his seatbelt. "Let's hope the engine doesn't blow up."

Twelve minutes later, tires squealing, we turned off the highway and onto Lochside Drive. Up ahead a police car was sitting behind three cars. We skidded to a stop beside the cop. "Don't worry," he said. "It's just a fender bender. Nothing more than a few bruises."

We put neck braces on two of the drivers and took them to Saanich Peninsula Hospital. After we arrived, I left our patients with Mark outside emergency and went up to the triage nurse. "Don't worry about them," I said. "They're fine. I'm the one with the problem. My nerves are shot and my stomach's on fire. Have you anything for acute indigestion?"

Really, they only left us a treasure map. What a hoax. I didn't treasure that call.

U-TURNS

We were sent routine to a minor accident on a main street in Victoria, and as we drove up, I saw a policeman standing beside two vehicles on the far side of the road. In order to park close to the accident, I needed to turn the ambulance around, so at the next set of lights I switched on our emergency lights, made a wide U-turn in the middle of the intersection and drove back up the other side of the street.

As I approached the accident, the cop charged over to our ambulance, shaking his fist angrily. He pounded on my door and motioned for me to lower my window.

Good grief, I thought. I've just used my emergency lights on a routine call and held up four lanes of rush-hour traffic while I pulled an illegal U-turn. He's going to give me a ticket, and my superiors will be getting a report! I rolled down my window apprehensively.

"That's unbelievable," the cop said. "Just outrageous! You had your lights on, and that blue car drove around you without even slowing down. If I'd been able to get a good look at his licence plate, I would throw the book at him. Some people have no respect for emergency vehicles!"

Merry Christmas

I've always liked to work on Christmas because paramedics are paid double time and a half, hospital wards are full of chocolate, and strangers are remarkably friendly. When we were called to a nursing home for a lady running a high fever, we were greeted by a frail choir backed by an equally elderly and tipsy band. Dominating the cheerful carolling were the grunts and wheezes of a very loud and off-key trombone. The manager shrugged and smiled. "That's our resident moose."

For most of us Christmas is a special time when families reconnect and reaffirm their love for each other: a time of good food, fun and snuggling up in front of a roaring fire. But for those unlucky enough to have lost their families because of misfortune, mistakes or migration, Christmas can be an especially lonely time.

THE BEST PRESENT

I was a lucky kid with a happy childhood. We lived outside of Ottawa with fields on one side of the house and the Rideau Canal on the other. The winters were cold, and most years our world was buried in deep, white snow.

On Christmas mornings my brothers and I would wake up early and empty our Christmas stockings. We would pull back the curtains and peek through the beautiful swirls of frost covering the windows, and even though it was still pitch-black outside with no sign of dawn, we would wake our parents, turn on the hall lights and run downstairs to the living room. Invariably, while we were sleeping, gifts would have appeared under the Christmas tree. They were wonderful presents like a red Montreal Canadiens jersey for my older brother, a blue Toronto Maple Leafs jersey for me, new skates and hockey sticks, and one year a long toboggan—perfect presents for three boys. We would help the other kids in the neighbourhood scrape the snow off the thick river ice and make a rink. Then we would play hockey every day after school and every weekend until our hands were too frozen to hold our sticks.

Of course, I wanted my two children, Lars and Zara, to enjoy Christmas as much as I had. Fortunately my brothers and my parents had followed me out to Victoria, so my whole family—including my ex-wife—could get together for birthdays and holidays. One year, we decided to have an early family celebration on Christmas Eve, since I would be working Christmas Day.

After we arrived at my parents' house, we brought out our gifts and piled them under a twinkling tree. We had all bought gifts for each other, and soon the presents formed a ridiculously large mound. Then, when we thought we were finally finished, my younger brother Duncan brought in an enormous box that was taller than my six-year-old son. The box was covered with beautiful green paper, tied with golden ribbon, and topped off with a lovely big bow. My children stared at it greedily. "Of course it's for the two of you," Duncan said to them, "but you have to open it last."

We had a delicious Christmas dinner, but the kids barely touched their food. They stole away from the table and began shaking the big green box in a vain attempt to discover what kind of treasure it held.

It took us an hour to open the other gifts. By then each child was surrounded by a heap of toys, clothes and torn paper. But they were far from satisfied. Ignoring their loot, they kept asking Duncan, "Can we open the big present now?"

Their uncle finally nodded, and Lars and Zara dragged the enormous package to the middle of the room, where they pulled off the ribbon and paper and opened the box. Inside the box was a slightly smaller box, this time wrapped in blue paper with a red ribbon and bow. Confused, they pulled out the blue box and tore it open, only to find a silver box inside it, this one lovingly tied with a black ribbon and bow. By now all the adults were laughing, and the kids, increasingly frustrated, yanked out the silver box and tore it apart. Once more they found an even smaller box, wrapped in bright orange paper. Finally they pulled out a small green box, tied like the first one with a golden ribbon. By this time my children were getting angry, and they quickly ripped it open. Inside were two grinning hand puppets. Lars and Zara dropped them on the carpet in dismay: they didn't know whether to laugh or cry.

That joke was the best gift of all. I doubt if my children can remember many of the presents they were given when they were little. But I bet they'll never forget that big green box.

CHRISTMAS DREAMS

The next morning it was Christmas Day, and I went to work with a head filled with sweet memories and a lunch bag stuffed with good food. But I was soon reminded that not everyone is as fortunate as my family.

We were sent to pick up a sick man in his early sixties—a double amputee. He lived in a tiny ground-floor apartment where his bed, refrigerator, sink, stove, couch, table, TV and wheelchair were all crammed into one room. On the table was a small Christmas tree with fake presents that he had bought from Canadian Tire.

I introduced myself and gave him a thorough examination. Although his vitals were normal and his lungs clear, he looked much older than his age: haggard and exhausted. "I don't know why I'm still living," he said in a barely audible voice. "What for?"

I leaned close to Mark and whispered, "He's not that sick. I think the real problem is that he's lonely and depressed. He called an ambulance, but can we take him to the ER if he just has a bad cold?"

Mark shrugged. "Let's give it a try. The poor guy desperately needs a change of atmosphere, even if it's just a trip to a waiting room."

The man told me his story on the way to the hospital. He had lost his legs at age eleven when he was walking across a railway bridge and got hit by a train; later he left Ontario and moved out to Victoria because he was tired of his electric scooter getting stuck in the snow. He had never married, and now that his sister and brother had died he didn't have any close family.

At emergency, I explained his situation to the middle-aged triage nurse, Marie. "He's not really physically or mentally ill. Can you do anything for an old guy who's feeling sad at Christmas?"

Marie winked. "I'm sure we can find some reason to admit him. We're not busy today and with his history he's bound to have medical problems." She walked over to the listless patient on our stretcher and gave him a big hug. "Merry Christmas! They tell me you need a proper tune-up along with a bit of tender loving care, so why don't we bring you inside? You're just in time for a nice Christmas meal!"

The man was surprised. "You're really going to look after me? And give me supper?" For the first time I saw him smile. By the time we lifted him onto an emergency room gurney, he was starting to perk up and had colour in his cheeks.

Mark and I were in a good mood when we climbed back in the ambulance. "Marie has real Christmas spirit," I said. "Nothing cures people like loving kindness. Now how do we get the health-care system—and for that matter, the whole world—to act like it's Christmas every day?"

Mark snorted. "You think governments will build houses for the homeless, and rich people will share their wealth with the poor and hungry? You're dreaming."

But, I thought, sometimes Christmas wishes are granted. If enough people dream of a better world, we can make these dreams come true.

By the Book

It was one of those shifts. We had gone from one call to the next without a break for six hours. At 9:00 p.m. I called the dispatcher from the hotline at the emergency room and asked whether Mark and I could grab a quick zero in the hospital cafeteria. "You can try to have supper," was the reply, "but I can't guarantee you'll eat much. Almost every available ambulance is tied up, so if anything happens I'll have to call you."

I had just set my plate in front of me and stood our black portable radio beside it when dispatch called: "Sixteen Charlie! Sixteen Charlie!"

I dropped my fork in disgust and picked up the radio: "One-Six Charlie, on portable."

"Sixteen Charlie, on the air for a Code 3, possible Code 4."

"Let's go, Mark," I said. "It sounds like someone's dying."

I ran out into the cool spring night and jumped into the ambulance, booking on the air and writing down the address and response number. Behind me, Mark walked calmly out to the car, climbed into the driver's side and carefully closed his door. As usual his cautious style annoyed me, clashing as it did with my more reckless pace. Oh, God! I thought. Now he'll crawl at sixty kilometres an hour across town. Our patient will be good and gorked by the time we get there. But Mark surprised me by immediately cranking the key, throwing on the emergency lights and sirens, and taking off at a good clip. The car was rocking around the first corner while I was still snapping on my seatbelt and slipping a pair of orange earmuffs over my head.

We were soon rolling down the patient's street, and I read out the house numbers as we went. "Six hundred—two blocks to go. Five hundred. Okay, now we're in the four hundred block and looking for 415. There's a fire truck ahead, that must be it. One minute, it's parked outside 450—maybe we're going to 450, not 415?"

Mark turned off the sirens but kept on driving. "No, they're just giving us plenty of room to park."

In the gloom no numbers were visible on the tall apartment building. I opened the window to ask directions from a person walking by, but he just kept on walking.

"Let's ask that lady over there," I said, and Mark drove across the parking lot and confirmed that we had arrived at 415. Then he turned the ambulance around, pulled up in front of the foyer and shut down the lights. Angry with myself for having delayed our arrival by questioning the address, I said, "Well, that stupid little search wasted another minute." After three to four minutes in cardiac arrest, brain cells begin to die.

We pulled out the stretcher and placed our jump kit and portable oxygen tank on top of the sheets, with Mark as always taking a few extra seconds to secure the tank with a seatbelt. "Don't bother getting our defibrillator out," I called to him. "Here comes ALS." A big Advanced Life Support ambulance was pulling into the apartment parking lot, its flickering strobe lights looking like an invitation to a gala event. We went on ahead of the other crew, breaking down the stretcher along the way to enable us to squeeze into the narrow elevator. On the way up I turned on the oxygen and checked the equipment.

A middle-aged fireman in a black uniform was waiting for us in the hallway. "This way, boys!" he said, waving. "But take your time—there's no need to rush."

After a dozen years as a paramedic, not much surprised me, but I hadn't expected to walk into such a tranquil scene. Two more firemen were having a comfortable chat in the brightly lit living room while other calm voices came from a bedroom down the hall.

"Where's the patient?" I said, hurrying forward.

"There." The younger of the two firemen pointed toward a high-backed yellow armchair. "She doesn't have a pulse."

The chair back hid an emaciated old lady with thin white hair. I got down on my knees to look at her. She was sitting up with her head leaning on her chest, dressed only in a shift. Expressionless eyes stared unblinkingly from a worn face. I took out my stethoscope and listened for a heartbeat. There was none, but her body still felt warm and pliable.

"Why haven't you started CPR?"

The younger fireman looked slightly uneasy. "Her husband says that she died at least fifteen minutes ago. He called his lawyer instead of calling us, and the lawyer's here now. He's the one who called us. Apparently she doesn't want to be revived—so we haven't done anything."

"Hmm." I looked at my partner. "That's fine by me. We're supposed to attempt resuscitation unless we have a legal order not to, but it doesn't make any sense in this case. After fifteen minutes she must be brain dead, so the kindest course is to just leave her in peace."

I stood just as the ALS crew walked in, looking cheerful and relaxed as usual. Frank, a white-haired paramedic who doubled as an evangelical preacher, was carrying the monitor. Eric, a former race car driver with a lined, friendly face and tattooed arms, was following with a large drug kit. "So, what's happening?"

We explained that we had decided not to attempt to revive her. "You're probably right," Frank said, "but to do things properly we should hook up the monitor and make sure there's no cardiac activity."

Reluctantly, I attached the leads while Frank switched on the machine. A green line etched its way across the screen, roughened in places like ripples on a pond. Frank smiled gently. "Look, it's in a fine V-fib [ventricular fibrillation]. Her heart is still struggling."

"Now don't start anything!" We all looked up to see a bent old man with dark pouches under his eyes standing in the hallway. "She's been sick for a decade, and for the last five years she's been in so much pain she prayed every night for death. Don't you bring her back just so she can suffer some more!" A man in a suit and tie, who I guessed to be his lawyer, put a hand on his shoulder, and the old man turned and shuffled back into the kitchen.

Frank called out, "One minute, please. Did she sign a Do Not Resuscitate order?"

"No," the lawyer replied. "I suggested she sign one, but she never did. But I can attest that she wouldn't have wanted to be revived."

"I'm sorry," Frank said, "but we have a legal obligation to attempt resuscitation unless our patient has signed a DNR form."

"Come on, Frank," I said, "you heard the husband. Let's leave her alone."

Frank shook his head. "No, that wouldn't be right. She didn't tell us not to do anything. We have to give her a chance—that's what we're paid for." He looked at Eric, the senior paramedic present.

Eric shrugged then nodded. "Let's go for it."

I pulled an oral airway, Ambu bag and face mask from our jump kit, and by the time I straightened up the others had opened the blanket pack on our stretcher and lifted the old lady from her chair onto the stretcher. She now lay on her back, staring at the ceiling. In a moment we had cut her nightshift to pieces, revealing a body that was nothing but bones covered with wrinkles, as if all the life in her had been used up.

I twisted the airway into her mouth and passed the bag and mask to the younger fireman, who began ventilating the woman. His partner started chest compressions while I connected oxygen tubing to the portable oxygen tank and hooked it up to the bag. Across from me Mark was setting up the intravenous equipment, Eric was getting ready to intubate, and Frank opened boxes of syringes filled with cardiac drugs.

"She's too thin; you'll never find a vein on her," commented one of the firemen. We all glanced at the woman's left arm, where Frank had fastened a tourniquet. A long blue-black vein was appearing on her forearm. The cardiac compressions were working.

"You're wrong," I said. "She has the bad luck to have good veins. Eric won't have a problem getting an IV in there." And in a moment he had. Mark hooked the tubing up to the needle and then went off with a clipboard in search of a medical history from the husband. I traded places with the fireman doing the compressions.

"Hyperextend the neck," I told the younger fireman. "You're getting too much air into the stomach. If you get rid of the pillow, you can do it better."

It didn't really matter because Eric was already moving around the stretcher to intubate. He pulled out the airway and inserted the laryngoscope into the woman's throat, bending her neck right back for easy visualization of the trachea. I noticed that she had no teeth. "Can't miss with this one," he said. "I can see clear through to the other end!" He inserted an endotracheal tube, attached the bag to it and checked to ensure that a good volume of oxygen was getting directly into the lungs.

Everything was going like clockwork. Across from me, Frank had put gel on the paddles and was ready to defibrillate. "Stand back!" he said, placing the paddles on the woman's chest and pressing the trigger. The body jumped with the shock and we all looked at the monitor. No change. The fireman and I went back to our CPR. The monitor whined as it recharged.

Frank called out again, "Stand clear!" and once more the frail body jerked. We looked at the monitor. This time large waves filled the screen. "We've got the beginnings of a good rhythm," he said. "Let's see if there's any output."

Eric felt for a pulse at the throat. "No pulses. Let's give her some drugs."

The fireman and I resumed CPR, but I felt uneasy. "Frank, if you bring her back as a vegetable, I'm holding you responsible. Don't try too hard."

He ignored me and began injecting drugs into the rubber port on the IV line.

Just then we heard the sound of the husband's laughter coming from the kitchen. The younger fireman was astonished. "That doesn't sound like a grieving husband. I wonder if he wanted her dead."

I couldn't raise a hand without breaking my stroke, so I nodded over my shoulder toward a shelf where a small framed picture showed a beaming husband and wife standing with their arms around each other. It must have been taken ten years earlier when they were both in their seventies and the woman was still healthy and plump. "They look like a loving couple to me," I said. "I'm sure he's simply relieved to see an end to her misery."

The monitor indicated that the woman's heart was weakening, and Frank asked Eric, "What do you think? One more epinephrine?"

"Yeah, give her another epi ..."

The old man came out of the kitchen again, followed by Mark and his lawyer. "She just wants to die," he said. "Five years ago when the pain started, we tried to commit suicide together, but it didn't work. Since then I've been nursing her every night, and now she can finally have some rest. Please, just let her go." Then he turned slowly and disappeared back into the kitchen.

On the monitor the green line had gone almost flat. Frank raised his eyebrows. "It looks pretty agonal. I guess we should stop now. We've given it a good try."

Eric nodded and reached for a telephone. "I'll get orders from an ERP to call it off." After explaining the situation to an emergency room physician, he received permission to stop further attempts at resuscitation.

"Okay, you guys," Eric said. "We can quit now."

I stopped compressions and straightened my aching back. "Good," I said, sighing. "Now we can leave her alone."

Everyone looked relieved, and we disconnected the monitor leads, the IV line and the ventilation bag, and covered the little body with a sheet. Then Eric asked dispatch to notify the police and the coroner's service, and we began cleaning up the empty syringes and scattered debris and packing up our equipment. When the husband returned, we told him that his wife was gone for good.

"I'm glad," he said. "This was what she wanted more than anything else." He looked fondly at his wife, lying pale and still on the stretcher. Apart from the plastic tube protruding from the corner of her mouth, she looked peaceful. He bent down and kissed her lips softly. "Goodbye, my darling. We had sixty good years together. Now I only hope that I can die soon too."

We walked out of the apartment, pulling the stretcher with us. I wished the husband well and suggested that he ask a close friend or relative to stay with him for the next few days.

"I'll be all right," he replied. "Nothing matters any more. Perhaps I'll join her in heaven."

Then Frank expressed his sympathies and we all filed out quietly into the hallway.

In the elevator I said, "Frank, what were you thinking? By the time we got there, she had been down for at least fifteen minutes. If you had revived her, she would have been a brain-dead turnip."

He winced. "Don't blame me—I don't write the laws. I was just doing my job."

As Mark and I drove away, I thought about the absurdity of what we had just done. There are times when we do wonders with our technology, but there are times when we overreach ourselves. Everybody talks about the sacredness of life, but whatever happened to the sacredness of death?

Spaced Out

I've worked with some brilliant paramedics—people who love medical challenges. In crises they spring into action, work quickly and flawlessly and perform miracles. But I'm happy if no one becomes sick or injured and I'm not needed. I have no objection at all to a quiet shift filled with long coffee breaks and pleasant conversations.

Nevertheless, most of the time I'm pretty focused when I work. Nothing concentrates your thoughts like driving a Code 3 through rush-hour traffic or trying to stop a patient from bleeding out on your stretcher. But sometimes the mind wanders.

EARTH TO GRAEME

The patient was complaining of chest pains. Because he had a history of angina and all the symptoms, after checking his vitals we helped him take a nitrate pill. As soon as his pain went away, he cheered up and decided he didn't need an ambulance—he had an appointment with his GP the next day and would be fine until then. All the same we insisted on taking him into emergency for a thorough checkup.

After we loaded him into the back, I radioed dispatch to let them know that we would be going Code 2 to the nearest hospital. It was a lovely, warm spring afternoon and I enjoyed the familiar drive down streets covered with pink cherry blossoms. We passed my kids' primary school. It had been completely rebuilt to make it earthquake-proof, and the architects had designed a beautiful building. Sometimes governments actually get it right. I thought about my kids. It was almost time for school to finish for the day, and they would be walking home soon.

A few minutes later I pulled to a stop, turned off the engine and opened my door. That was when I heard a voice from the back. "Earth to Graeme! Aren't we supposed to be going to the hospital?"

Good Lord, I was on autopilot! I had forgotten I was in an ambulance and had driven home for lunch.

After we admitted the patient to emergency, I took my partner aside. "I'm really embarrassed about this ..."

"Don't worry," Don said. "I once drove a patient with a fractured leg to the nearest coffee shop! If you don't say anything, I won't either."

OLD TRICKS

That wasn't my only mistake. I kept losing my glasses.

The worst calls were cardiac arrests. On one call we found an elderly man lying on his back on the living-room couch. He had no pulse and no respirations, so my partner and I immediately moved him to the floor and began CPR. As I knelt down, I took off my glasses and put them on a nearby coffee table. (I need glasses for driving, but close up I see better without them.)

After a few minutes an ALS crew showed up, hooked up their monitor and started an IV. For the next half hour we were all busy co-ordinating compressions and ventilations, intubating the patient and checking the monitor, running ECG strips, infusing drugs and taking blood pressures. The man's heart began to beat and we stopped CPR.

Once his pulse and blood pressure were stable, the ALS crew was ready to leave for the hospital with the patient. "Is the stretcher ready? Okay, take off the leads and move the tubing out of the way. On the count of three, lift him onto the stretcher." One of the crew strapped down the oxygen bottle and hooked the monitor on the sidebar. I put the laryngoscope back in the drug box, closed the lid and passed it to Andy, the senior ALS crew member.

"Graeme, you come with us," he said. "Mark, you can clean up and follow. Tell the family to take their time—there's no point getting in an accident on the way to the hospital. Their dad's doing fine."

I reached for my glasses. No glasses. No coffee table! Everything had been moved in the mad rush to look after the patient. "Excuse me. Has anyone seen a pair of glasses?" The family ignored me. They were hugging each other and sobbing with relief.

"Where did you put them?" my partner asked. "Why don't you keep them on your head?"

I finally found them. The coffee table had been put behind the couch, and my glasses were underneath it on the floor.

I ran down the stairs and out to the ambulance. Andy was busy piggybacking a drug bag onto the IV line. He glanced at me as I climbed in the back door. "Did you get lost? Next time remember to look for a big truck with red flashing lights."

Eventually I learned to put my glasses in my shirt pocket. That solved the problem of losing them at work.

But it didn't solve everything. At home I don't wear shirts with pockets and I lose my glasses at least once a day. My wife Ferie is trying to train me to always leave them in the same place, but no luck. How does that saying go? Something about new dogs and old tricks.

COLD COMFORT

If I wasn't perfect, neither was anyone else.

The elderly woman called us because she was "feeling poorly." She suffered from half a dozen degenerative diseases, including severe osteoarthritis. While I assessed her, Rick wandered around the house collecting information and filling out the form. He reported that the refrigerator held more medications than food.

The woman was weak and could walk only with difficulty, but this wasn't surprising as it looked as though she hadn't had a proper meal in days. Possibly she was also overmedicated. Clearly she needed a thorough medical examination along with some tender loving care, so we lifted her onto our stretcher, tucked her in and headed for the door.

"Hang on a minute," Rick said, patting his pockets. "I've lost my stethoscope!" This was serious. You can't play medic without a stethoscope, and they don't come cheap.

He looked all over the apartment while I chatted with the patient. Nothing. "How could it just disappear?" he said. "I had it around my neck."

"Retrace your footsteps and you'll find it," I advised.

So Rick started in the hall, went to the bedroom, then to the living room and finally to the kitchen. As a last resort he opened the fridge and took another look at the meds. His stethoscope was there, next to a jar of strawberry jam. "Damn," he said. "I'm losing it."

I didn't say a word as we carried the patient down the stairs. But I felt relieved, even happy. It was a great feeling to know that I was not the only space case in the Ambulance Service.

That Glowing Feeling

We were doing a routine transfer from the main part of the hospital around the block to the psych wing. As we were loading the elderly woman into the back of the ambulance, I noticed a line of food services staff pushing loaded steel carts down the ramp from the new kitchen. They all wore the same blue hairnets and sterile starched uniforms and walked like a row of robots.

I grinned across the stretcher at my partner. "The way those food workers are dressed, you would think they're handling nuclear waste!"

We drove off, and on the brief trip I did my best to reassure our patient that the psychiatric hospital was the best place for her to deal with her fears and delusions. I've always prided myself on my ability to calm disturbed patients, even those who are completely off the wall, so I was pleased that, by the time we arrived, our patient was resting quietly. But when we opened the back doors and lifted her out on the stretcher, she let out a wild yell and started thrashing around.

"I know what you want to do! You're trying to kill me!" She was grabbing at the straps, trying to unbuckle them. "You'll never get me in there! Let me go!"

"What's the matter?" I asked, confused by the sudden outburst. "Nobody is going to hurt you. We're all here to help."

"Stop lying to me!" She was furious, and I turned to see what she was looking at. It was one of the food service workers pushing a cart into the psychiatric wing. "You're not fooling me!" our patient yelled. "You're not taking me in there and stuffing me full of radioactive waste!"

Break and Enter

In emergencies, paramedics encounter all kinds of bizarre situations, meet all kinds of eccentric people and get to do all kinds of forbidden things like cutting off people's underwear or driving over carefully manicured greens at golf courses. But sometimes simply getting to the patient offers unexpected challenges, such as the time we were asked to check up on an elderly woman who had seemed unusually confused when her daughter called her from Calgary.

The old lady answered when I buzzed the apartment intercom. "Who's there?"

"We're from the Ambulance Service. Can you open the door?"

"Who?"

"The Ambulance Service. Your daughter wants us to make sure that you're all right."

"I can't let you in—I don't know you."

I said louder, "Please open up. We're from the Ambulance Service."

"I don't know you."

"We're paramedics. We're here to help you."

"You're who?"

"We're the police."

"Oh, why didn't you say so? Come on up."

BREAK A LEG

The woman was in her eighties, and she knew she was in trouble when she tripped on the edge of her living-room carpet and fell on her oak floor. The loud crack and sharp pain when she hit the hardwood told her that she had broken her hip. In agony she dragged herself to the corner table, pulled the phone to the floor and called 911.

Five minutes later Susan and I arrived outside her apartment door and knocked. A shaky voice responded from the other side, "The door is locked, and I can't get to it to open it."

I called through the door, "Does anyone else have a key?"

"My son does, but it would take him an hour to get here. I can't wait that long—I can't take the pain!"

Susan said, "If you want, we can break in, but it's going to damage your door."

"Go ahead! Do whatever you have to do. Just hurry up!"

I looked at Susan questioningly. Because doors are designed to keep people out, they don't fracture easily. I remembered going to an apartment with a fire captain and his crew. As we had to break in, he asked for a volunteer to smash the door. The firefighter who volunteered was a powerfully built young man—the handsome, cheerful type you find on firemen's calendars—and he confidently backed up across the hall and launched himself against the door. He struck it with a loud thud and staggered back, clutching a dislocated shoulder: the door had a metal core. Another time I had watched while a policeman tried to kick in a wooden door. He hit it so hard his foot punched through to the other side, leaving him hanging from the door by one leg.

So I said cautiously, "Are you sure we can break it down? Maybe we should get the cops to bring a battering ram."

"Come on, we don't need any help!" Susan replied. "Do you want to do it or shall I?"

Now, Susan is not your average woman. A former member of the provincial judo team, she's tall and muscular, and I don't think she holds wimpy males in high regard. She's married to Neal, a big ALS paramedic, and rumour has it that she fell in love when a psychotic patient attacked her with a knife, and Neal decked him with one blow. No, there was no way out—I would have to prove I was a real man even if I broke a leg trying.

"I'll do it," I said. Long ago I had studied karate (for two months), so I turned sideways and aimed my foot above the lock. As I did so, I thought, "I'm going to regret this!" but I let go with all my strength. To my shock the door completely shattered, with fragments flying deep inside the apartment and the handle and lock falling to the floor. As

Susan and I stared, the door frame followed, collapsing one jamb at a time.

She gave me a big smile. "Wow! That was amazing!"

I have to admit that was the macho highlight of my life. Of course I was lucky to be up against a really cheap door: it must have been made of balsa wood.

OVER THE WALL

It was a lovely summer day in Victoria when Mark and I were called for a fall in a kitchen. We arrived at a pretty blue house, carried the stretcher up the wooden steps to the porch, set it down under tinkling wind chimes and rang the doorbell. There was no answer. In the middle of the front door was a large stained-glass window, so I peered inside through a small rose pane. "I can't see anyone," I said.

Mark tried the door and checked out the windows. They were all locked. "We might have to break in."

"I'd hate to wreck anything," I said. "It's obviously a well-loved home. Let's try the back."

We went around the side. Behind the house the garden was enclosed by a tall brick wall. I reached up and pulled myself high enough to look over the top. "There's a big yard with lots of flowers. Hey, the back door is open and I can see legs! Someone's lying just inside. Help me get up—I'll go in through the back and let you in the front door."

Mark boosted me up to the top of the wall. Then he passed me the jump kit and I slid down the wall into the yard. I was quite pleased with myself. There, I'd made it without even getting dirt on my uniform! I turned around just in time to see a huge black Alsatian galloping toward me.

I grabbed my jump kit and held it in front of me. "Stop!" I shouted. "Stay!"

But that big Alsatian was in no mood to put up with a home invasion. Thousands of years of breeding as a guard dog were in his soul, and he wasn't going to miss this opportunity to savage an intruder. He dodged back and forth, barking, snapping and growling. "Nice doggie," I pleaded in as calm a voice as I could muster. "Lie down." The animal paid no attention—he was searching for a way to tear me to pieces.

I was desperate. I thought of throwing the dog my stethoscope, but I didn't think he would take the bait. So I backed along the flower bed, jerking the jump bag up, down and sideways for protection. Finally I made it to the back door, squeezed inside and slammed the door, locking it as well just in case the dog turned out to have an above average IQ.

Then I examined our patient and, still sweating, opened the front door to let Mark in.

As they say, always look *very* carefully before you leap.

Where's George?

Mark hung up the hotline. "What an exciting day! First we have a kid with a stomach ache, then two angioplasty transfers, and now we get to pick up a patient with dementia. Apparently she drank a bottle of bath oil."

"Why not?" I said. "It's probably a great cure for constipation."

The patient lived in a one-bedroom apartment on the second floor of a townhouse in a seniors' complex, and after we'd carried our heavy stretcher to the top of the stairs, I grunted, "I can't believe these buildings don't have elevators. Old people live here!"

Mark shrugged. "It was probably designed by a hundred-year-old Swedish fitness freak. Did you know that people who have to walk up and down a flight of stairs every day live an average of five years longer than their ground-floor neighbours?"

"Thank you for that information," I said. "With all the exercise we get in this job, we should make it to a hundred and fifty."

A home-care nurse was waiting for us. She explained that although the patient had mild dementia, she was normally able to take care of herself with a little assistance. "However, she doesn't usually drink bath oil! My main concern now is that she's running a fever, which is making her dementia worse."

Mark read the ingredients on the back of the empty bottle. "I don't think this stuff's toxic, but we'll take it with us to emerg. They can decide whether she needs her stomach pumped."

"Hello!" I greeted the patient cheerily. "How are you feeling?"

"Fine. Never felt better."

That didn't sound right, so I tried a test question: "Do you have any children?"

"One or two."

"Do they visit you?"

"Sure! Several times an hour."

Mark grinned. "I'd say she's missing a few parts of the puzzle."

The nurse said, "Ann, you have a fever, so these men are going to take you to see a doctor at the hospital."

"I can't go out like this! Give me my makeup!"

The nurse found the patient's wig and lipstick, and after we lifted her onto our stretcher, I took her vitals: "Pulse 160, temperature 40."

Mark, who was bent over the stretcher tucking in the sheets, glanced up and said, "Forty? She could be delirious."

Ann chortled. "Are you calling me dearie? Oh, you're so cute. Give me a kiss!" She grabbed Mark around the neck and planted a ring of red lipstick on his cheek. He pulled back in surprise.

Laughing, the nurse turned to the patient. "Ann, dear, you know you should always wear clean underwear when you go to the doctor." She looked under the sheets. "Oh, she doesn't have any on."

"That's okay," Mark said. "They'd just have to take them off in the hospital. She's not going dancing."

The old lady brightened. "We're going dancing? Whoopee! It's going to be so much fun. Where's my pink dress?"

The nurse patted her arm. "Don't worry, you look just fine, dear. Now you'd better get going. You don't want to be late."

"Is George coming with us?"

"Who's George?" I asked the nurse.

"Her younger brother. He died three years ago."

Mark and I wheeled the stretcher out of the apartment. After retracting the undercarriage at the top of the stairs, Mark picked up the foot end of the stretcher and I took the head end, and we started down the steps. Halfway down, the old lady suddenly reached out and grabbed the bannister. Thrown off balance, we jerked to a stop.

"What are you doing?" she demanded.

"We're taking you to the hospital."

"No! I don't want to go to the hospital."

"You need to go. You've been doing poorly."

"You're poor? Then why don't you get a job?" She was angry now. "There's no excuse for a healthy young man like you to sit around all day. And take your friend with you. Lazy layabouts. You're disgusting. And where's George? I'm not going anywhere without George!"

"We're sorry," I said, "but George can't come. He died three years ago."

"You're lying! George is younger than me."

We were in a quandary. We were stuck halfway down the stairs with the stretcher in mid-air. Neither of us could free a hand without dropping our patient. "Mark," I said, "pull gently and maybe she'll let go." But the old lady hung on to the railing with an iron grip.

"George, is that you?" she said, addressing Mark. "Take off that beard, George, and stop hiding from me. And stop hiding my dolls!"

I had an idea. "Mark, try biting her foot. If she's distracted, maybe she'll let go."

"I don't remember reading that in the manual, but here goes ..." Mark leaned toward the stretcher and gently bit the sheets covering the old lady's right foot.

"Ow!" she yelled, kicking at him. "Who did that?" But she still clung to the railing.

"That was Graeme," Mark said. "I'll give you his payroll number if you want to file a complaint."

"We'll have to try something else," I said, and I blew on her face.

"Stop that, you rude man!" she yelled and slapped me with her free hand.

I blew harder, and this time she got really mad, letting go of the railing as she tried to squirm out of her seatbelts. We quickly carried her down the rest of the steps, lowered the undercarriage and wheeled her over to the ambulance. To our surprise as soon as we were outside, her mood changed completely, and she started singing, "Row, row, row your boat, gently down the stream ..."

I shook my head in amazement. "With the raging fever she's got, she should be sleeping, not singing!"

Two kids were standing on the sidewalk next to the ambulance, and one of them asked, "Where are you taking her?"

The old lady said, "We're going to a party. Do you want to come?" She began waving her arms as if she were conducting a band.

By the time we rolled our patient into the emergency room, she had her wig on backwards and lipstick smeared all over her face. At triage the nurse looked up from her desk and said, "Good Lord. What's this?"

Wait for the Paramedics!

We were sent routine for a woman who had fallen on Government Street. When we arrived, I was surprised to see a middle-aged woman lying face down on the pavement with two other women sitting on her back.

"Why are you sitting on her?" I asked as we approached.

"She's our friend," one of the sitters replied, as if this gave them permission. "She tripped and fell on the road. We keep telling her it's not safe for her to move until she's been checked out by the paramedics, but she keeps trying to get up. So we're holding her down."

"Can you get off her now?" I asked, and the two women got up. I leaned over their friend. "Do you hurt anywhere?"

"No—apart from feeling squashed, nothing hurts."

"Well then," I said, "why don't you stand up?" I helped her to her feet.

The poor woman was furious. "I've being saying that nothing is wrong with me, but they wouldn't listen. They kept repeating, 'Wait for the paramedics! It says on TV that you have to wait for the paramedics!' What nincompoops!"

They say real friends have your back, but I don't think they're supposed to take this literally.

Drug ODs

One day Mark and I found ourselves standing on a wind-swept wharf in Victoria, looking at the distant line of snow-capped mountains that mark the American side of the Strait of Juan de Fuca. We had been dispatched to meet a police boat that had rescued a cocaine addict who was trying to swim to Port Angeles.

I said, "He probably thinks it's cheaper to buy coke in the US."

Mark laughed. "I guess he hasn't heard about the exchange rates."

Although we couldn't resist the occasional joke, drug addiction isn't very funny. Susan and I had attended an addict who had taken too much heroin and was about to go into respiratory arrest. While Susan was manually ventilating him to keep him alive, and I was filling a syringe with medication to reverse the overdose, the man's stoned girlfriend staggered over and began patting his pockets. "Do you think he's got some cigarettes?" she said. "I've gotta have a smoke." What an ugly scene: a damaged, addicted couple living in hopelessness, squalor and despair.

Sometimes drug addicts are lucky and everything goes right; sometimes their luck runs out. I guess it helps to stay on the right side of the angels.

HE MUST HAVE A GUARDIAN ANGEL

We were only two blocks away when we got the call for a Code 3, possible Code 4. Mark stepped on the gas and in no time a frantic bystander waved us to a stop. A car was stalled in the middle of an intersection. Inside, a stout middle-aged man in short sleeves and a tie sat motionless behind the steering wheel, staring into space.

I jumped out of the ambulance, ran over and checked for respirations. "He's not breathing," I yelled at Mark. I felt for a carotid pulse. "Weak, irregular pulse." Hmm—that was odd. Then I looked at his eyes. "Pinpoint pupils," I said. "He's overdosed!"

Mark passed me a large plastic airway and I twisted it into position in the man's mouth. Then Mark handed me an Ambu bag, and

I began pumping air into the man's lungs while Mark connected an oxygen tube.

"Yeah, he's got track marks on his arm," Mark said. He pointed, "and there's a syringe on the floor. He must have just shot up. I'll get some Narcan."

Two minutes later Mark injected the medicine, and almost immediately the man began to breathe on his own. His pulse quickened and he regained consciousness. We lifted him onto our stretcher and then into the back of the ambulance.

He looked around in confusion. "What's going on? What happened?"

"You nearly died from a drug overdose—that's what happened."

"I did?"

"Yes, you moron! If we'd arrived a few minutes later, you would have been gone for good. You better get on your knees and thank God you're still alive. Next time you might not be so lucky." I didn't usually speak to patients so harshly, but he really had me worried.

"Oh no, I can't die! I've got a wife and two kids. That's the last time I'll ever do smack. I swear I'm going to quit!"

Did he? I hope so, but I just don't know. We never saw him again.

OUT OF LUCK

The call came in as a Code 3 for a possible shortness of breath. We arrived at the apartment a few seconds before the ALS crew, so I rang the doorbell. Nothing happened. I rang again. Through the small glass panel I could see a young woman walking slowly toward the door.

I turned to Mark, who was standing in the hall beside our stretcher. "I guess there's no rush." Behind us the ALS crew—Michelle and Josh—emerged from the elevator with a defibrillator and drug box. The woman unlocked the door and motioned us inside. Without saying anything, she turned and wandered down the hallway past two doors to a living room where two men and a woman were sitting on a pair of matching couches. Our guide sat down beside them. Still no one spoke a word.

"Is someone short of breath?" I asked.

One of them pointed his thumb over his shoulder. I stepped closer and looked behind his couch. Another man about twenty years old was lying on the floor, his face turned to the wall.

I squeezed past the couch and felt for a breath. Slow and shallow. Then I took his pulse: weak and irregular. I pulled out my flashlight and looked at his pupils: tiny and non-reactive. "Hey, you guys! Give me a hand—this fellow doesn't look too good. He may have OD'd. Let's get him on the stretcher and start bagging."

Mark and I lifted him over the couch and onto our cot while Michelle turned on our oxygen tank and connected it with a tube to an Ambu bag and face mask. In the meantime, Josh quizzed the others in the room before turning back to us.

"Yeah," he said. "They've been shooting heroin. They're all stoned."

I placed the face mask over the man's mouth and nose and began squeezing the Ambu bag to force oxygen into his lungs while Mark moved the stretcher to the hallway to give us more room. While Josh attached the monitor leads, Michelle drew up a dose of Narcan and injected it. "He should be coming around in a minute," she said cheerfully.

But nothing happened. Josh exchanged a look with Michelle. "Let's give him another dose." Done. But still no response. I stopped bagging for a moment and listened for respirations.

"He's stopped breathing," I said, and resumed bagging.

Michelle said, "I'll intubate. Mark, how's his blood pressure doing?"

"Systolic down to 60."

Josh had readied a bag of saline on the IV pole and was probing with a catheter for a vein. "All his veins are flat. Not normal for a young guy." He moved around the stretcher to try the other arm.

Michelle had inserted the intubation tube, and she put her stethoscope on his chest while I connected the ambu bag and pushed a little air down the tube. The man's stomach rose.

"Damn!" she said. "It went down the esophagus. I could have sworn I had it right." She pulled out the tube. I ventilated for a minute and she tried again.

"How's the IV coming?"

"No luck. I'm going to have to put a line in his jugular."

Mark had been watching the monitor. "He's in arrest. I'll start compressions."

The intubation tube was finally in place, and now his chest moved easily every time I squeezed the bag. Michelle groaned. "Man, that was difficult! What's wrong with this guy? Josh, as backup I'll do a cutdown on his foot. We've got to get an IV going right away." She dug into the drug box and pulled out a scalpel.

"Don't bother," Josh said. "I've got jugular access. Now let's see if we can restart his heart."

Half an hour later the floor was littered with syringes and empty drug boxes, but despite our best efforts, the line on the monitor stayed flat.

From behind us a shaky voice asked, "Will he be all right?" It was the woman who had let us in.

"No, he won't be all right!" Josh was angry. "He's gone. Dead as a doornail."

Our patient was young and healthy, and we'd reached him before he stopped breathing. He should have survived, but he didn't. For some reason he ran out of luck.

A Drive in the Country

We were sent routine for a seventy-nine-year-old man suffering from shortness of breath. His doctor had called dispatch to have him transported to hospital and admitted for pneumonia. It was a beautiful summer evening and we were happy to go for a drive in the country. Mark was at the wheel and I was riding shotgun.

Soon we were out of town and past the streetlights. Stars were breaking out in the darkening sky and Mark started naming them: "And that bright one right next to Ursa Major is Arcturus. It's twenty-five times the diameter of our sun ..." I nodded to hide my ignorance. Mark could have talked all night about the stars—he collects telescopes and should have been a science teacher—but he must have noticed the clueless expression on my face and politely changed the subject.

At the bottom of a long hill two flashlights waved us down, and Mark pulled up in front of a neatly trimmed hedge. The patient's middle-aged son and daughter led us through a floral arch and down a long path to a pretty house. We carried our equipment inside and upstairs to a bedroom. Everything looked tidy and immaculately clean.

The old man was obviously a devoted husband: his wife had been a quadriplegic for twenty years following a bad car accident, and since then he had looked after her every need, including changing her urine bag and diapers. Normally he slept beside his wife, but he had moved to the spare room after he became ill so she wouldn't catch his bugs.

I opened his shirt to listen to his lungs with my stethoscope. Under his white chest hairs was a crude drawing of a heart pierced by an arrow. "Did you get this when you first met your wife?"

"Hell, no," he said between coughs. "I got roaring drunk the day I got out of the Navy and some bum tattooed me while I was passed out."

He shivered as we lifted him onto the stretcher and wrapped him in the bedding. "I don't know why I feel so cold. Can you warm me up?"

Mark said, “We could raise your fever, but perhaps for now we’ll just add another blanket.”

I strapped the oxygen bottle next to his legs, slipped a cannula into his nose and turned on the tank. We wheeled him over to the next room and lowered the cot so he could say goodbye to his wife. When he saw her, tears came to his eyes. “I’m going to miss you every minute,” he said, sniffling.

She smiled. “Don’t worry about anything. Alice will stay with me until you get back.”

We carried him down the steps and out into the warm, jasmine-scented night air. The son guided us back to the road, quietly warning us to watch out for the rose bushes crowding the path. The daughter walked beside her father, holding his hand and stroking his hair.

It was a magical moment. When we reached the trellis at the entrance, it felt as though we were passing under a triumphal arch: a testament to this family’s love and devotion. People can have such strange fates, with happiness and pain all mixed together. And sometimes tragic events turn ordinary people into saints.

Irregular Shifts

Days unfold randomly for paramedics. Every shift, we were sent to places we had never been before to meet people we had never met before, often to deal with completely new situations. Some of these can be very challenging, but others are just strange.

For example, one afternoon we were sent to the Coast Guard helipad to pick up a man suffering from severe hypothermia. It had been his first attempt at windsurfing and no one had taught him how to tack. Six hours later a Coast Guard helicopter spotted him heading out to sea. By the time they handed him over to us he was so cold and rigid he could hardly move his lips.

While the work was usually interesting, it wasn't always a walk in the park. Sometimes it got pretty stressful.

IN THE LINE OF DUTY

It was while backing up police SWAT teams that I really learned to respect cops and follow their orders. Their first rule was to stay far away from windows and doors whenever there was an armed suspect inside a building, so my partner and I would stand well back while police in bulletproof vests would creep along the walls, shout to the suspect, and if he didn't surrender, smash down the door with a metal battering ram and rush inside. That takes guts.

One pleasant summer day Mark and I were sent down to James Bay to back up police who were responding to an incident involving guns. We drove slowly up the street looking for a police car. "There's not a cop in sight," I said. "They must either be inside the apartment building or around the back. I guess we'd better stop and look around."

I pulled up in front of the building. Then I heard a voice say, "Hey, what are you doing?"

I looked across the street and suddenly realized that a half-dozen cops were crouching in a line behind the parked cars.

"Get out of here!" the voice said. "You're right between us and the shooter!"

We got out of there fast.

LINE DANCING

The dispatcher called me at home on my last day off. "Graeme, you're IV certified, aren't you?"

"Uh, yeah," I said, confused by the question.

"Good. I need you to drive an ALS car next block. You'll be working with Pierre on 29 Alpha, starting tomorrow at 7:00 a.m."

"But I've never worked on an Advanced Life Support car before," I protested. "Of course I've assisted ALS hundreds of times, but I don't know their protocols or drugs."

"Pierre will tell you what to do. Just follow his directions." And *click*—he hung up.

I had known that sooner or later I would be asked to fill in on an Advanced Life Support ambulance, but I'd always thought that I would be given some training first. Now I felt as if I were being thrown to the wolves, though at least in this case the wolf was Pierre, and I trusted him. He was the kind of guy who would have enjoyed working as a trauma surgeon in a war zone. He loved the challenge of emergency medicine, and nothing made him happier than a real disaster with multiple critical injuries. Then he was in his element, rapidly triaging patients and making life-saving interventions. If you ever find yourself in a plane crash, believe me, you want someone like Pierre to show up and save you.

The next morning I got to Station 29 forty-five minutes early. Luckily the night shift was still in the station, so I was able to climb into the back of the big ALS ambulance and start examining the drugs, equipment and other supplies—hundreds of items. I focused on the heart monitor, the intubation kit and the large portable drug box with its bags of intravenous fluids, IV catheters and tubing, boxes and vials of drugs, anaesthetic sprays, defibrillator gels and monitor patches.

As I tried desperately to remember which drugs were in the red, blue, yellow, green or black boxes and what was in the various vials, I felt as if I were cramming for an exam.

That was when the side door of the ambulance opened and Pierre stepped in. "Hi, Graeme! They told me that you were working with me this block. Great!"

"Don't be too happy," I said. "This will be my first shift on an ALS car."

He laughed. "Don't worry. I'll tell you what to do. With a bit of luck we'll get busy and have lots of fun." He passed a couple of monitor batteries to me. "These are fully charged. Keep an eye on the battery level and change batteries if the charge gets too low. Also check the printer tape often. Don't let it run out." He ran a practised eye over the drug box and the cabinets before saying, "Back in a minute!"

I looked at my watch: quarter to seven. I should have at least fifteen more minutes to cram. But then the hotline rang, and a minute later Pierre climbed into the front passenger's seat. He turned around, passed back a handful of drugs and said, "It's Code 3 for chest pain. I told the other crew that since we're both here they can relax and go home. Okay, get in the driver's seat. Time to rock and roll!"

So began one of the most stressful days of my life. For the next six hours critical calls came in one after the other without a break: serious heart problems, a paralyzing stroke, an antidepressant overdose and an almost fatal case of internal bleeding. At each call Pierre would rattle off orders such as, "Let's get her on the stretcher. Now get a complete set of vitals. I'll attach the monitor leads and run a strip. She's hypotensive so put her head down and give her high-flow oxygen. Start an IV with saline. Okay, another set of vitals—but this time just blood pressure and pulse. Did you write them down with the time? Good, don't forget to keep records. Now piggyback this drug on the bag of saline and run it at fifty drops per minute. Take the clipboard and get the patient information and collect her meds. I'll run another strip and get the firemen to carry the stretcher down the stairs. While we're taking her to the ambulance, you can clean up and pack up the drug box. Be quick. Then you can get another set of vitals while I telephone the ER. Then drive us there Code 3."

Each time the emergency room staff would be waiting for us, and as soon as we arrived, we would be ushered to a monitor bed. I would help move the patient, disconnect our equipment and put our stretcher, monitor and oxygen bottle back in the ALS ambulance. Then, while Pierre was giving his report and filling out his forms, I would clean the equipment, put fresh linen on the stretcher, restock the drug box, take out the garbage and mop the floor of the ambulance. Or at least I would make a start on these jobs, but every time Pierre would come rushing out before I was finished. "Don't bother with the floor, Graeme. We've got another call!"

We finally got a break at 1:00 p.m., which allowed Pierre to catch up on his paperwork and me to finally clean the back of the ambulance and reorganize the supplies. I had just finished when he came out of emergency with a big grin. "What a great morning! Are you having fun?"

I groaned. "Not yet. It's a lot to learn in a short time."

He clapped me on the back. "All you need is more practice. We've got forty-two hours to go in this block, so with a bit of luck we'll get lots of good calls."

And we did. Most of our patients had critical problems, and I had to work as fast as I could while triple checking to make sure that I was drawing up the right drug and administering it at the correct dosage.

I'll never forget one of our patients: a man in his early twenties who was having an acute asthma attack. Although he had been using his inhalers, he could barely breathe by the time we arrived. He couldn't talk either, but we could see from his terrified look that he was suffocating. I immediately tried to assist his ventilation with a bag and mask, but his airways were too constricted. Pierre did a quick assessment and said, "I'll have to intubate him, and while I'm doing that, you start an IV so I can give him epinephrine."

I connected intravenous tubing to a bag of saline and ran the fluid through the line. Two firemen had joined us, and while one of them helped Pierre, I asked the other one to hold the IV bag and tubing. Next, I inserted a large needle into the young man's arm. It slid easily into a vein, and I connected the tubing to the catheter, taped everything in place and checked to make sure the fluid was running slowly but steadily.

By now Pierre had successfully intubated our patient and connected the end of the plastic tube to our Ambu bag and oxygen. He asked the fireman beside him to start ventilating and then turned to me. "Is the IV ready?"

"Yes," I replied. "Wait a minute—it's stopped running!" Blood was starting to back up the transparent intravenous tubing.

Pierre wasn't pleased. "Well, hurry up and fix it."

This was no time to screw up! What was I doing wrong? I was sure I had put the needle in the middle of the vein. I checked the catheter site: everything looked fine. I looked at the bag of saline: nothing wrong there. Then I started tracing the thin plastic tube back from the patient's arm. The IV line ran down to the floor—to where the fireman was standing on it. "Can you move your foot?" I said.

"Oh, sorry!" The fireman almost danced in embarrassment. As soon as he lifted his foot, the line began to clear and immediately Pierre bent down and gave the first drug injection. I've rarely felt so relieved.

At the end of the block Pierre said, "You did well! Have you thought about taking Advanced Life Support training?"

"I've thought about it," I said, "but because I'm a single parent now, I don't think I could find the time to study."

But I was lying. Other paramedics who were parents managed to find the time to study. The truth is that I've never been that interested in learning about diseases and drugs.

My paramedic friend Dan once said (half-joking), "This would be a great job if it wasn't for the people!" But that was because he liked the medical side best. Me, I liked working with people—the happy ones, the sad ones, the ones who were delusional, the ones who were afraid and suffering. I always hoped to be able to help at least some of them.

I probably would have made a better paramedic if I were more like Pierre, but I've never enjoyed crises. My idea of a good ambulance shift was one where no one got sick or injured. Too bad that only happened in my dreams.

Complimentary Services

Marv and I were about to lift a big woman out of her wheelchair and onto our stretcher when she said, "Be careful. I'm heavier than I look."

I said, "No, you're not. Uh, I mean yes you are..."

Marv shook his head in disbelief. "You sure know how to compliment a lady!"

One Step from Eternity

In this business, we talk to a lot of dying people and we see a lot of death. It comes in all forms, sometimes quickly, sometimes slowly. Some people die peacefully, some die in pain or fear. I've had a middle-aged man fall into my arms, his eyes wide with surprise as a massive heart attack robbed him of life, and I've revived an elderly man once, only to have him cry out for his mother and arrest again, terror frozen on his face. Others have been glad to die, eager to leave old pains and an old body behind and move on.

But at the end of our days death waits for all of us, and I find it both fascinating and a privilege to be with people who are making that mysterious transition. Having responsibility for another's life or death makes medical work an incredibly serious affair, but also changes its nature from routine to awesome. There are many times when one is overwhelmed by the pain and tragedy of a situation, but also frequent moments of wonderful spiritual depth. I have learned much, particularly from those who approach death with acceptance and dignity.

Some of my clearest memories are of people who were alive and then dead, as well as people who were dead and then alive, but one of the most pleasant is of the man who died twice and didn't know it.

THE MAN WHO DIED TWICE

My partner and I had arrived at the house a few minutes before the Advanced Life Support crew, so by the time Michelle and Josh walked in the door, we had already put the elderly cardiac patient on oxygen and loaded him onto our stretcher. We then pulled the stretcher out of the hallway and into the living room to have enough room to hook up the monitor and start an intravenous line.

Arriving close behind the ALS crew was the man's doctor, an older gentleman who was obviously a personal friend of the family. As soon as the patient spotted the doctor, he began to joke about going on vacation to the hospital. Suddenly he broke off and fell silent.

I felt for a carotid pulse and said, "I think he's in arrest."

The patient's wife put her hands to her face in horror.

"Okay," Michelle said calmly. "You guys start CPR, and we'll intubate and take care of the IV."

We put in an airway and began to manually ventilate the patient and compress his heart while the ALS crew prepared their equipment. When the patient gagged, we quickly stopped CPR and removed the airway. He opened his eyes and smiled. "Oh, sorry," he said. "I must have fallen asleep."

The doctor laughed. "Yes, you must have. Rather rude of you to go to sleep in the middle of a conversation, but I'm not surprised that you're feeling a little tired."

I noticed the wife relax a little.

Josh inserted a needle into the man's left arm and started an IV. Michelle checked the jagged line on the monitor screen. "It's V-tach," she said. "Let's start with some Lidocaine." She injected a syringe into the intravenous line. Suddenly the man looked fixedly at the ceiling. We all glanced at the monitor, where the line had gone flat. "Commence CPR. He's arrested again."

The doctor was watching silently. Behind him, the wife sagged against the arm of a chair.

Michelle pressed the defibrillator charge button on the monitor and began smearing gel on the paddles while Josh lined himself up at the patient's head with a laryngoscope. "Stop bagging," he ordered, and pulling out the airway, he tilted the patient's head back in preparation for inserting the scope.

Just then the patient opened his eyes again. "Are you all still here?" he asked. "Hmm. I seem to be passing out."

All eyes were on the beeping monitor. "Good. He's back in a sinus rhythm." The doctor smiled cheerfully at his friend. "So how is your chest pain now?"

"Well, I can't say it's worse, but it somehow seems different. The deep pain has gone, but now I feel really sore. It's like an elephant has been jumping on me."

We all laughed.

"You should start feeling better now," Michelle reassured him. "Your heart has stabilized. It's beating slower and stronger." From the top of the IV pole, a second bag was steadily dripping more drugs into the patient's vein.

The doctor turned to the wife, who by now had slumped into the chair. "And how are you doing?"

"I don't know … I can't really say …" Her voice was shaking. "This is all a bit much."

"Well," said the doctor authoritatively, "what you need is a cup of tea. Your husband is in good hands. Let's go into the kitchen and chat."

"You're absolutely right," she replied firmly as she stood. "What we need is a strong cup of tea. Would you like a little cake to go with it?"

STILL HOPING

The nurse asked whether I could stick around for a couple of minutes. "My patient has terminal cancer. The docs think he won't last more than a few days. He's pretty fragile—just skin and bone—so it will save him a lot of pain if you can lift him up while I change his bedding."

We walked into the private room where an emaciated body lay hooked up to IV lines and drainage bags. The nurse touched his arm lightly and a pair of pale blue eyes snapped open. "Maureen! What a pleasure to see you! You've never looked lovelier. Come sit by my side and talk with me!"

"Mr. Phillips, we're here to change your bed," she told him. "This kind paramedic has offered to help me."

"Ah, we don't need him. All I want is your sweet company and your delicious touch!" He put his frail hand on top of hers. Maureen deftly slipped to the side, folded up his bedcovers and placed them on a chair.

After moving the tubes out of the way, I slid my arms underneath him and gently picked him up. The old man kept watching the nurse, who was smoothing the sheets with well-practised efficiency.

As soon as I laid him back on the bed, he said, "Darling, could you please fluff up my pillow?"

When the nurse leaned over to raise his head, the old man quickly reached up, taking a breast in each hand. "Just perfect!" he said, sighing.

Maureen froze, carefully disengaged his hands, kissed him on the forehead and straightened up. "Goodbye, Mr. Phillips. I'll check up on you soon."

We left the room, closed the door and burst out laughing.

"I thought you said he was almost dead."

Maureen nodded. "He is. But with some men the testosterone goes last!"

DYING FOR A SMOKE

We were dispatched Code 2 for a sixty-nine-year-old male with end-stage emphysema. As I expected, he was wheezing with every breath, his skin was grey from lack of oxygen and he was utterly exhausted. But to my surprise, he was not lying in bed propped up on pillows but sitting upright on a straight-backed wooden chair.

"Glad ... you're here," he gasped. "Haven't slept ... for four nights. Can't breathe ... if I lean back. Usually tough ... it out. Not this time."

I changed his nasal cannula to a Venturi mask to help his breathing, though I doubted this would do much more than show him that we cared, and switched the tubing over to our portable oxygen. While Mark was writing down his medications, I recorded his vital signs and asked him about the history of his disease.

I said, "You don't just have puffers [bronchodilators] within reach; you also keep packs of cigarettes on hand. Shouldn't you give up smoking?"

"Oh, no," he said between coughs. "I love smoking!"

"How long have you been smoking?"

"Since I was fifteen ... two or three packs ... every day."

"Tell me, would you have started if you knew how much suffering it would cause you?"

"Oh, sure ... I love smoking."

We bundled him up and delivered him to the hospital, probably for the last time.

Afterward we went back to the ambulance and remained parked outside emergency to wait for our next call. "What a miserable way to go," I said, "suffocating to death."

"Yeah," Mark agreed. "It's an awfully slow way to commit suicide. But you heard him—it's his choice."

"On the way over here he told me he used to be president of a big pharmaceutical company. So he certainly knew the risks."

Mark moved the driver's seat back to make room for his long legs. "We're all addicts," he said. "Some people eat too much, some drink too much, others watch too much television. Who are we to judge? That man smoked even though he knew it would shorten his life, and that was his right."

I nodded, and as we sat in silence for a few minutes, I found myself thinking about a refugee I had met in Toronto. "I knew a Chilean guy called Rafael," I told Mark. "He was a union organizer, and when the military overthrew the government, he was arrested along with tens of thousands of other people and held in the National Stadium in Santiago. They tortured him for several weeks, and then took him outside to be executed. But just before they were about to shoot him, he asked the officer in charge of the firing squad whether he could have a last request."

Mark was intrigued. "Really? What did he want?"

"Rafael asked if he could have a cigarette. He said, 'I promised my wife I'd give up smoking, but now I don't have to worry about dying from cancer.' The lieutenant started laughing. 'That's funny,' he said. 'You're a brave man. Not many people can make jokes when they're about to die. I won't kill you today. Hey, José! Take this man back inside and bring me another prisoner. We'll shoot someone else.' Rafael doesn't know why, but a few months later the military let him go, and he was able to make it to Canada. So smoking doesn't always kill you."

"Right," said Mark. "And a couple of people have fallen out of planes and lived because they landed on haystacks. But I'm not going to try it!"

THE OLD LEOPARD

The front door had been left ajar, so we pushed it open and walked inside, pulling the stretcher between us. In the living room an ancient man was sitting in an armchair in a dressing gown, surrounded by the mementos of a full life: rows of thick books, tall blue vases, richly woven rugs and inlaid tables covered with black-and-white photos of long-gone people and places.

As we walked in, the old man looked up sharply. "Who are you? Where are Lucy and Diane?"

"They're off today," I said. "You'll have to put up with us."

"What a pity. They're so much prettier than you, and since I can't pick up girls anymore, it's nice having them pick me up."

"So what's bothering you today?" I asked.

"There's so much to say I wouldn't know where to stop," he said.

"You mean start."

"That too."

"Take your time," I told him. "We're not in a rush—we're paid biweekly. Do you hurt anywhere?"

"Are you kidding? I'm ninety-six—you should ask me where it doesn't hurt. My eyes hurt, my neck hurts, my back hurts, my feet hurt—but that's not why I called you. I've got congestive heart failure, and today I'm just too tired to get up. I need to go to the hospital to get my drugs adjusted."

"Okay," I said, "but first let me take your blood pressure and check your breathing while my partner writes down your medications. Hey, what happened to your arm? It's covered with scars."

"Oh, that happened when I was young and wild. One night I got so drunk that I walked into my neighbour's house by mistake. His dog took offence and bit me. I bit it back, but as you can see, I lost the argument."

"So you had a wasted youth?"

"No, I had a great time—much better than now. Now I pass my days with my mates—all guys like me with shiny false teeth. Our average age is ninety so we say we all have long-life disease, the disease of dying slowly. With time passing I don't dare look in a mirror. I grew up in Persia, and in the old days I would look in the mirror and say in my language, 'What a good-looking guy! The reason God made you so beautiful was to drive women crazy!' Now I speak perfect English, but when I look in the mirror all I can say is 'Shit!' My black hair is gone and my bald head flashes in the sun. At least I don't have to spend money going to the barber. Now I'm covered with yellow and brown spots. I have so many that I look like cheap wallpaper. My mates call me the Leopard."

Mark came out of the bedroom with the clipboard and a handful of pill bottles. "You are Mr. Ali Amiri, I presume?"

"Not Mister—Professor. I used to teach international law."

"Okay, Professor. What's your birthday?"

"February 12."

"But what year?"

The old man winked at me. "This guy's not too bright. Every year, of course!"

Mark grinned. "Well, if you're ninety-six, I guess I can figure out when you were born ..."

I put my stethoscope back in its holster. "Professor, I can hear some crackles in your lower lungs, but otherwise your breathing isn't bad. You're probably right that your CHF is making you tired. Are your legs more swollen than usual?"

"A little," he said. "They also feel colder and more numb."

Fortunately his condition wasn't critical, so we didn't have to whistle him to the hospital, and we invited him to keep telling us his story while we got the stretcher ready. The professor was happy to have an audience.

"Now I spend more time in hospital than in my own house," he said. "I go there so often I'm thinking about renting an apartment next door. Last year I went to the hospital so many times for my stomach, my back, my knees, my eyes, my feet and my heart that I now know the doctors better than my kids. I have had so many endoscopies, colonoscopies, MRIS, CAT scans and X-rays that my hospital photo album is fatter than my family album. Soon I'll glow in the dark like a ghost!

"I used to be proud of my round bum, but it's become as flat as the bottom of a flowerpot. And every year I have to go to my tailor to have my pants shortened. And I'm getting thinner and thinner. Instead of using a belt, I keep my pants from falling down with a rope—it's easier to tighten. And because I don't want people to think I'm totally deaf, I spent $3,200 to buy the smallest hearing aid available. It was so tiny it got lost in my ear and I had to pay a specialist another $200 to get it out! I've given up trying to understand people when they talk to me. I used to just say 'yes,' but today I mostly nod.

"In the old days I used to work out at the gym, but I've run out of energy so now I only do half the bench presses—the bench part! I'm like a newborn baby now. I don't have hair, I babble instead of talk, I can't walk and I wet myself all the time. A couple of days ago I went to the urologist and told him that I'm having more and more trouble peeing. He said, 'How old are you?' When I told him ninety-six, he said, 'Then don't worry about it. You've peed enough in your life already.' Big help he was. It's got so bad I don't send my pee sample to the labs: I just send my pants."

By now Mark and I were in stitches. "Come on, you're making all this up."

"I'm not making up anything! You'll see if you get to live to my age. It's not much fun when your warranty runs out, but as long as you can laugh, you can manage."

Mark said, "We're ready to go, Mr. Leopard. Let's lift you out of your chair and onto our stretcher."

"Not Mister—Professor!"

"Okay, Professor Leopard."

In emergency, the triage nurse took one look and said, "Oh God, him again! Isn't he dead yet?"

"That's not very kind," I protested. "He's a lovely old fellow."

"Maybe, but he's at the stage where we can't do anything for him, and he knows it. He just comes here to chat with the staff. He should be in an extended-care hospital, but he refuses to leave his house and his memories. He says everything there reminds him of his wife, and it would kill him to move. Whatever. I'd like to leave him in the hallway, but he's a cardiac patient so he gets to go to the front of the line. Go on then—put him in bed one."

A NEAR-DEATH EXPERIENCE

We were given a routine call to take an elderly diabetic female to the hospital for the amputation of a gangrenous foot. Mark and I carried the stretcher up three flights of stairs to an apartment on the top floor of what had once been a stately mansion. There we were met by her older sister.

"You'll find my sister in the bedroom. She's not hard to find—she's hardly moved in fifteen years."

I helped Mark put our patient on our cot. Then, to protect her blackened toes from the weight of the sheets, he began making a bridge out of some flexible orange splints. While he was busy, I went to the kitchen to get some information from her sister.

"Can you tell me her medical history?" I said.

"Of course! I know everything about her—I'm her nurse, nanny, cook and cleaner. To make it short, she's eighty-eight years old, and you name it, she's got it. Because she has a strong will, nothing can kill her and she keeps on going. But each year she gets worse and complains more."

"I'm sorry," I said. "The hospital will need her complete medical history and all the medications she's taking."

"They should have all that. She's been there often enough—it's the high point of her life when she gets to leave her bedroom and visit emergency."

She opened the fridge door and pulled out dozens of pill bottles. "Here, fill your boots." For the next ten minutes I copied the information onto my clipboard, writing in tiny letters to fit everything onto the one-page form while pressing hard to make four copies. While I wrote, Mark wheeled the stretcher out into the hallway, chatting cheerfully with the patient.

I tucked the clipboard under the mattress at the head of the stretcher and Mark slung the jump bag onto his back. Then we began carrying the patient down the stairs. I took the top end and Mark walked backwards, holding the foot of the stretcher above his head to keep the patient from sliding under the straps and hitting her toes. When we reached our ambulance, we lifted the stretcher into place, Mark jumped inside and I closed the back doors.

I asked the older sister whether she would like to ride up front with me and opened the passenger's door. "Let me help you up," I said. "This is a truck, not a car, so it's a bit of a climb."

But she waved me aside—"I'll be fine, young man!"—and nimbly hopped into the cab. I was astonished. Pretty good for a ninety-year-old.

I went around the car and got into the driver's seat. On the way to the hospital, I remarked, "Your sister may have a few problems, but you're in amazingly good shape."

"Don't you believe it. I died last year. I was in cardiac arrest, but you paramedics revived me. You broke two ribs."

"My apologies. Sometimes it's difficult to know how hard to push when we're doing compressions." Then I had a thought. It was a long shot, but why not ask? "Did you happen to have a near-death experience while your heart was stopped?"

"Yes, I did," she replied. "It was just like you read about in the magazines. I felt peaceful, like all my troubles were over, and I started travelling down a long tunnel toward a beautiful, warm light. I knew my family would be waiting for me, and when I reached the light, my mother and father, my brothers and all my friends were there. I was so happy! Then my mother told me, 'You can't stay. You have to go back.' I said, 'Why? I belong here!' But my mother said, 'Not yet. You have to go back and look after your sister.' Next thing it felt like someone was punching me in the chest, and I woke up to see you guys."

"Well," I said, "the good part is that you get to be here a bit longer."

"What's good about it? I hate looking after my sister!"

TOUGH CHOICES

Jason and Al were a good team. During the three years they worked together they had become best friends—they were even golf partners on their days off.

On one shift they took a comatose patient home from the hospital, where she had been treated for an infected feeding tube. A year earlier this eighty-seven-year-old woman had suffered a massive stroke, and her medical records stated that there was no chance of her ever recovering consciousness. For the last month her middle-aged daughter had been looking after her with home-care assistance, and it was this daughter who opened the door when they arrived and helped the paramedics put the patient into bed. Then she lifted the gown covering her mother's frail body. "Look at these bedsores! This is why I don't want her to stay in the hospital. But her health is slowly deteriorating, so no matter what we do, every week the sores get worse. You wouldn't let your dog die like this."

Jason nodded sympathetically. "It's a crazy system. Health care should be about helping terminal patients have good deaths, not forcing them to stay alive and suffer."

The daughter looked at him hopefully. “My brothers and I would like our mother to have a quick, peaceful death. Her GP agrees—he has offered to get the drugs and load the syringe, but he’s not willing to risk his licence by ending her life. And much as we think it’s what our mother would want us to do, no one in the family can push the plunger. Please,” she begged, “will you do this for us?”

Jason looked thoughtfully at the unconscious woman, the tubes snaking from under the bedcovers and the suction equipment lying ready on the bedside tray. Her breathing was laboured and noisy, and her arms were extended and rotated, a clear sign of a severe brain injury. He looked back at the daughter’s tired, worried face. “Okay,” he said. “I’ll drop by after work.”

His partner was shocked. “What? Are you crazy? Euthanasia is illegal. You could lose your job over this. And I could lose mine for not reporting you.”

Jason looked at him steadily. “Sometimes we have to stop passing the buck and do what’s right.”

“Oh no,” Al snarled, “don’t get all moral on me! In fact, don’t say another word. Don’t tell me what you’re going to do. I don’t want to know anything about it.” Angrily he snapped the sidebars up on the stretcher, threw the O_2 bottle on top and pushed it outside to the hallway, slamming the apartment door behind him.

Jason joined him in the ambulance a few minutes later. For the rest of the shift they worked without talking to each other.

Two weeks later the woman’s daughter spotted them pulling up outside a streetside café. She walked over, put a hand on Jason’s arm, smiled and said, “Thank you from the bottom of my heart. You’re a brave guy.”

Al’s eyes narrowed but he kept his mouth shut. He went inside the shop and ordered two coffees to go.

THE OLD MAN AND HIS GRANDDAUGHTER

The man was in his eighties, very weak and short of breath, but he called out a friendly greeting as we lifted the stretcher into his house.

His heart was failing, and he had the worst edema I have ever seen: his legs had ballooned out to an enormous size, and even as we watched we could see the swelling creeping up his abdomen.

We whistled him to the hospital, and he was immediately given a bed in emergency. There the doctor checked him over and told him the blunt truth: "I'm afraid you don't have long to live. If we do nothing, you won't live more than an hour or two. Even if we do everything, you probably won't live much longer."

The old man was unconcerned. "I've been expecting this. Don't bother with more drugs. I'd just like to spend the rest of my time with my granddaughter."

He had phoned her at the same time he called for an ambulance, and she was standing outside in the waiting room. As soon as she came into the room, his face lit up with love, and he grasped her hand happily. I pulled the curtain to give them privacy and walked away to prepare for our next call.

I will never forget that last glimpse of the old man sitting propped up in bed, calmly waiting for death, full of smiles at the pleasure of his granddaughter's company. I hope that when death comes to me, I will meet it with as much peace and as little fear as he did.

Our Business Is Picking Up

Marv pressed the intercom at the rest home and announced, "Ambulance Service."

A staff member answered and asked, "Are you bringing someone back?"

"We're here to take out and deliver."

The voice at the other end laughed. "Like a pizza service?"

"Sometimes they're in pizzas."

Taking It Like a Man

Because Debbie had been off for two months with a back injury, when she reported back to work she was told to spend the first block riding third. During this time she was only to do light duties and not lift any patients. Mark and I were pleased to have her with us for four shifts as we enjoyed her company and were more than happy to give her all the paperwork.

On our second day we received a Code 3 for a brittle diabetic. At the house a worried woman in her mid-thirties said that after her husband Ed had injected his insulin, he got busy answering phone calls and forgot to take his afternoon snack. Now he was hypoglycaemic.

She led us to the living room where a stocky man sat on a couch. He stared at us with a confused and angry expression on his face.

"I tried to give him some sugar, but he won't take it. I'm scared to go near him. Last time this happened he gave me a black eye!"

"Do you have any orange juice?" I asked. "I'll give it a shot."

She poured a glass of orange juice, and I slowly approached her husband with it, talking calmly. "Here, Ed, have a drink."

His hand flew up and slapped my arm.

Although I managed to hang on to the glass, the juice sprayed over the rug. "Sorry about the mess," I said. "Mark, do you think we can get him onto the stretcher?"

"No problem." Mark is tall and strong as a horse. He positioned himself behind the couch, and I placed myself in front. Debbie lowered the stretcher, unbuckled the straps and pushed it closer.

"Now!" I said. Mark grabbed Ed's arms from behind and I grabbed his legs. We lifted him onto the stretcher and cinched the seatbelts tightly around his chest and thighs. Then Mark quickly changed positions. He moved to one side, leaned over the patient and pinned him down by his right wrist and left shoulder. Ed writhed, but he wasn't going anywhere in that hold.

"Okay, Debbie, you should be able to get a line into his right arm."

I plugged IV tubing into a bag of D10 [10 per cent sugar] and ran it through the line. In the meantime Debbie knelt opposite Mark, tied a tourniquet below Ed's right elbow and looked for a good vein.

Suddenly Mark let out a bellow. Unfortunately, Ed could still move his left forearm and he had reached up, grabbed Mark's crotch, squeezed and hung on.

"Aargghh!" Mark was turning red, but he refused to loosen his grip on the patient. "Debbie," he gasped, "hurry up with the line!"

But Debbie was shaking with laughter. Several times she tried to position the catheter, but the needle wouldn't stay still. Tears were rolling down her cheeks, and I wasn't in any better shape.

"Debbie, for God's sake, do it!" Mark begged.

Debbie bit her lip, looked at the floor and took some deep breaths. Her hand steadied and she slid in the needle. I hooked up the IV line and turned it on full.

Almost immediately Ed began to calm down. As he regained full consciousness, he released his grip on Mark and looked around. "Oh! Paramedics! Did I forget to eat again? I hope I didn't cause too much trouble."

Debbie smiled. "No, no trouble at all. I really enjoyed meeting you."

Mark said, "It was definitely interesting, but let's not do this too often."

Ed didn't want us to take him to emergency, so we pushed the empty stretcher back to the car. Mark was walking bowlegged. "I need an ice pack!"

"Aw, Mark," Debbie said, "you were a real hero. I'll write management and tell them that you refused to abandon your post even when injured and under attack. They'll probably give you a medal."

"What kind of medal? The Purple Nut?"

Veterans

Two crews were sitting in the station waiting for the hotline to ring, and somehow Eric starting talking about his days as a medic in the Army Reserves.

"We had a live ammunition exercise, and to add realism the instructor wired electrodes to a large container of gasoline. Then the company was ordered to charge up a hill, and just as the soldiers were topping the ridge, he set the thing off. There was a huge explosion, rocks flying everywhere, and all the soldiers hit the dirt. And the new radio operator yelled into his mike, 'Holy fuck! Over!'"

We all laughed and Mark said, "Canada needs an army. It's someplace to put homicidal personalities."

Of course, we can joke about our military, but without their sacrifices our lives would be miserable: if the fascists had won World War II they would have enslaved the world. I was fortunate—because the majority of ambulance patients are elderly people, I met many Armed Forces veterans and heard many amazing stories.

STILL FIGHTING

The old man had no feet, so we lifted him out of his wheelchair and made him comfortable on our stretcher. We were taking him to the hospital for dialysis, and on the way I asked him why both his legs had been amputated below his knees. "Did they have to take them off because of your diabetes?"

"No," he said, "I lost them during the war. I've been in a wheelchair ever since." He then told me about his unusual life.

He had been a gunner in a bomber in the last months of the Second World War. On their return from a mission over Germany they were attacked by fighters. Their plane was crippled, but they managed to make it back over Allied lines before bailing out. Although he had been wounded in both legs in the attack, he managed to crawl out of the plane and parachute safely to the ground. There he was picked up and taken to a British military hospital.

At the hospital the airman was seen by a surgeon who told him that, though the injuries to his left foot would heal, his lower right leg could not be salvaged and would have to be amputated. Two hours later they took him into the operating room and anesthetized him. When he awoke, he was shocked to discover that they had amputated the wrong leg.

The next day the surgeon dropped by the ward on his rounds, examined his legs and ordered more surgery. Then they wheeled him back into the operating room and removed the other leg.

"That's terrible!" I said. "They ruined your life. Aren't you bitter?"

"Why?" the old man said. "A war was going on. The hospital was full of wounded and the staff were frantically busy. Someone made a mistake, but it wasn't done on purpose."

"Did the government give you help after you got back to Canada?"

"Not much. When we came back, we were just told to get on with it. Also, the Department of Veterans Affairs wouldn't give me a pension because the mistake was made by a British hospital."

"So," I said, "first you lost your legs and now your kidneys are failing and you need your blood cleaned three times a week. That's really tough."

The old man looked up at me from under bushy white eyebrows. "No one ever said that life was easy. I was nineteen when I lost my legs, and I thought I'd never get a job and no girl would ever look at me. But my family helped me find a job, I met a lovely woman and we've been married almost sixty years. I consider myself lucky—most of my wartime friends were killed. I wake up every day glad to be alive. I've had a good life and I'm not giving up now."

THE ENGLISH OFFICER

An embossed brass plate on the front door read COLONEL ROBERTSON. We rang the doorbell and were led inside by a voluble housekeeper who explained that her employer had been retired for twenty years. Since his wife had died, he had been living alone, and although the housekeeper was there on weekdays and home care was usually available at other times, he really needed to be in a nursing home. "But he won't go, and when the Colonel says no, that's the end of the discussion."

He was sitting in a chair beside a desk piled with books and papers. His back was hunched and his head, which had only a smattering of sparse white hair, was bent forward over badly twisted hands that rested on the brown blanket covering his knees. Although he looked exhausted, he raised his head and smiled when we came in.

"Thanks for coming, boys! I'll be grateful if you can take me to see my specialist. My arthritis is giving me a bad time, and I can't move without help."

I knelt beside him to take his vitals and pulled up the sleeve of his silk dressing gown. To my surprise his whole arm was pitted and scarred from old wounds, and a long, deep scar was etched on each hand. "What happened to you?" I asked. "It looks like you fell into a threshing machine!" Yes, I can be a bit tactless at times.

He glanced up at me with blue eyes that looked surprisingly youthful in such a deeply lined face. "I was in the British Army in Burma during the war, and the Japanese captured us behind their lines. They treated us as spies. I was beaten for three days, but when I wouldn't tell them anything other than my name and rank, their commander had me dragged outside and staked to a log with bayonets through my hands. They left me there in the sun.

"They thought I'd break, but the torture just made me angry. I told myself that I'm an Englishman, and no English officer gives in to a Jap, and then to keep my mind off the pain, I recited Shakespeare out loud."

"You recited Shakespeare?"

"Yes. I studied English literature at Oxford before the war and did some acting, so I know my Shakespeare."

He continued. "The Japanese officers had been watching me, and on the second day the major came out and told his men to pull out the bayonets. Then he bowed to me, told me I was a real Samurai warrior, and sent me to their field hospital. So I survived, but my health never fully recovered. Although the broken bones healed, ever since the war I've had a difficult time with arthritis."

"And what did you do after the war?"

"I stayed in the army. I spent most of my career in Intelligence at NATO headquarters in Europe. Eventually I retired with the rank of colonel. Then I moved here to be closer to my daughter."

Mark and I lifted the old man onto our stretcher and tried to make him comfortable, but every position we put him in gave him pain. Finally he shook his head. "Don't waste your time, boys. Just get me to my doctor."

In the ambulance I said, "Not giving in was a moral victory, but your body is always hurting. Was it worth it?"

The old soldier smiled. "We won, didn't we? We had to beat them and we did. Of course it was worth it!"

BITTER MEMORIES

Not all the veterans we met were proud of their military service. Once, while we were taking an old man back to the veteran's hospital, I asked him whether he had been in active combat. He said he had fought in the Canadian Army in the Second World War, and it was horrible.

"Don't let anyone tell you that war is glorious. It's hell. When I joined up, I was a happy kid, as naive as they come, but nice enough. I'd help every old lady I saw. But the war made me a murderer and a cynic. I don't talk about it much because I have nothing but bitter memories of exhaustion, fear and death.

"Right after we landed in Sicily my best friend was killed while he was praying. And you know who killed him? An old woman with a grenade. After that my squad didn't care anymore, and we shot a young German soldier while he was surrendering. The rule was a good German was a dead German. I didn't try to stop anyone. I was scared and I didn't want to be next."

THE GERMAN SOLDIER

Another of my patients was a former German soldier who had also seen combat in Italy. One morning he was crossing a snow-covered field by himself when he was spotted by a British fighter pilot. The pilot flew low to get a good look at him—it was obvious that he was German from his helmet and uniform—and then the fighter banked, circled around and flew directly at him. There was no place to hide so he stood still, waiting to be shot. But at the last moment the British pilot pulled up, wagged his wings and flew away.

"I don't know what happened. Maybe his machine guns jammed, but I think he was tired of killing and decided to let me go. Amazing luck, eh? It's why I'm here today. I was tired of all the killing too—I never wanted to fight, but I was drafted—but for me the war went on for another two months until I was captured."

I asked, "If you didn't want to fight, why didn't you desert?"

"It wasn't possible. The Nazis had patrols behind the front lines looking for deserters, and you would be executed if you were caught. But the real deterrent was the Gestapo: if someone ran away, they would arrest his family. So we fought to the end.

"But again I was lucky. After my unit surrendered, I was sent to a POW camp in Alberta, and I just loved it. We were treated kindly, and after everything I had been through, it was paradise—never in my life had I seen such beautiful farms and forests. When the war ended, we were sent back to Germany, and I got married. But there was nothing for us there—the whole country was rubble. So as soon as we could, we immigrated to Canada. Best thing I ever did. Pretty lucky, eh? That's what I tell my kids: I have a wonderful family and live in a great country. I'm one lucky guy!"

I looked down at his fat medical file. He still had problems. But I had to agree with him. He had found happiness. We should all be so lucky.

THE MEDIC

Dr. O'Connell was an elderly ex-surgeon and an alcoholic. We'd get called to pick him up after one of his binges, and we'd find him lying in a ditch or on the floor in a cheap rooming house, filthy, incontinent and suffering from severe tremors.

The first time I met him, I saw that his right hand was badly misshapen from an old injury. "How did this happen?" I asked.

"I was in a North Korean prisoner-of-war camp and our side dropped a bomb just outside the wire."

"You were in the Korean War?"

"I was a medic with the paratroops. It wasn't nice what we did or what they did. When I was caught, I was tortured. I gave lots of information—I had no choice. Then I sold my body for rice so I could survive and give a little food to my mates."

"What a terrible experience!"

"Yes, but it taught me compassion. After the war I finished med school and then spent most of my career treating terminally ill cancer patients."

"Do your memories give you nightmares?"

"That's why I drink. It helps me forget."

When we got to emergency, I repeated his sad story to the triage nurse.

She laughed. "Don't believe a word. O'Connell's a drunk. It's all Irish blarney!"

THE TEDDY BEAR MAN

In my twenty-one-year career as a paramedic I went into thousands of homes, and while I have forgotten most of them, my memory of one particular apartment is as sharp as if I had seen it yesterday.

When my partner Mark and I knocked on the door, it was answered by a woman in her fifties who introduced herself as our patient's daughter. She ushered us into a bright, high-ceilinged hallway. Sunlight streamed in from tall windows on the left, the wall to the right was lined with photos and paintings, and the runner on the floor had an intricate Persian design. The effect was delightful. We followed the daughter down the hall to a large living room, which was filled with gold brocade upholstered chairs and couches and lined with antique walnut side tables and cabinets.

In the middle of this elegant room a white-haired man was sitting in an armchair. He was dressed in a light-blue bathrobe with matching slippers. The scene would have been charming if he hadn't been clutching a white teddy bear in his arms and staring into space with a confused, anxious look.

"Daddy," the woman said, "it's time to go. I can't look after you any longer. These men are here to take you to the hospital. They'll take good care of you there." Her father didn't seem to hear her. He kept twisting the ears of his bear.

She started to cry. "This is so hard. He lived here with my mother for thirty years. I'm grateful that she died four years ago because she didn't have to watch him deteriorate from Alzheimer's. Most of his life he worked as a bank manager, but he was much more than that." She nodded toward a row of black-and-white photos on the nearby wall of World War II planes and pointed to one of two young men standing beside a two-engined plane with Royal Air Force markings. "This is my father here on the left. He was a pilot."

Mark said, "That's a Mosquito fighter-bomber, isn't it?"

"Yes, he was an ace. He was given the Distinguished Flying Cross for shooting down three German planes and for following one of their new jets back to its hidden base and destroying another three planes in their hangars." She went to stand behind her father and put her hands on his shoulders. "To see him now you would never guess that he's a real hero."

Mark and I looked at the row of medals on the mantel, including a small silver cross. Then we lifted the old man onto our stretcher and took him and his teddy bear to an extended-care facility. In the ambulance I sat beside him, thinking about how we must all age with time. We rush through our lives, usually ignoring the elderly people we pass on the street. We have no idea who they were—or what we owe them.

What a Gas!

Kids' reactions to accidents and injuries are sometimes mystifying. Mark and I were on a routine call for a leg injury at Oak Bay High School and were directed to the middle of a playing field, where perhaps a hundred students were gathered. We piled our equipment on our stretcher and lifted and dragged it across the grass. A teacher called out for the kids to make way, and we went to the centre of the crowd where a fifteen-year-old boy lay moaning on the ground.

I knelt beside him and asked him what had happened. I checked his vitals and gave him a head-to-toe physical. "I'm going to have to cut your pant leg to look at the injury," I said.

But as soon as I touched his lower left leg, the boy yelled in pain, "Ow, ow, ow! Stop, you're killing me!" At this some of the other students gasped, and I could see some of them were crying in sympathy.

I exposed the injury as carefully as I could and found a fractured tibia. I turned to Mark. "This is going to need a tension splint. Better give him some Entonox."

Mark unstrapped the Entonox bottle from the stretcher and gave the boy the mask. "When you suck on this, it will help relieve the pain," he said. "But we can't administer it—you have to hold it yourself. Now take a couple of deep breaths."

The kid pressed the mask against his face and sucked in the gas. Then he giggled. "*Wooo!* This stuff's great!"

The kids surrounding us also started to laugh.

"Hang on now," I warned him. "It will hurt a bit when I take off your shoes and socks." I was trying to be as gentle as possible, but he screamed all the same. Again many of the kids surrounding us gasped and cried.

"Keep sucking on that gas," Mark said as he removed the traction splint from its bag and put it between the boy's legs.

To distract the boy I said, "Who were you playing against? It looks like you got hit by an elephant."

The kid started laughing. "No, I wasn't hit by an elephant. I was hit by a hippo, a big fat hippo! Hahaha!"

In sync the other kids also laughed. "He was hit by a hippo! Hahaha!"

We padded and splinted the leg, lifted the boy onto our stretcher, then asked for help carrying it, and the entire crowd trailed behind us while we packed the kid off the field. As we lifted him into the ambulance, they clapped, laughed and cheered.

"I know Entonox is called laughing gas," I said to Mark, "but I never knew a broken leg could be this funny. If I could only make my jokes work this well ..."

A Sticky Business

The man yelped when I stuck a large needle into a vein on his forearm. "Ow! That hurt."

My partner Marv remarked, "You have to excuse him. He's paid to be a prick."

Dead on Arrival

Someone had called 911 after realizing that a person sitting behind a window hadn't moved for two days. When Hal and I arrived at the house, we weren't surprised to find that the man was not only dead but that he was starting to smell a bit ripe. Before paramedics pronounce someone dead, we are supposed to place a stethoscope on the chest and spend at least two minutes listening for a heartbeat and then three minutes listening for any sounds of breathing. But this rule doesn't apply if there is lividity, rigor mortis or signs of decomposition, and in this case the body was clearly past the stale date. So we called for the coroner, Code X'd and drove off in search of a coffee.

As I turned the first corner, I said, "The neighbours didn't look very happy when we left without a patient."

Hal grimaced. "Yeah. People really expect us to revive every dead body." He paused before adding, "You know, it's funny how people react to death. Sometimes it bothers them, sometimes it doesn't."

"You're right," I agreed. "Last October I responded to a guy hit by a bus. He burst open like a ripe tomato and there were splatter marks everywhere. A crowd had gathered and was staring at the scene, and none of them seemed particularly upset. Of course, it's different when someone close to you dies, but even in those situations people can surprise you. I once told an old woman that her husband had passed on. Instead of breaking down in tears, she said, 'Thank God!' It turned out they'd had a terrible marriage, but she had stayed with him because she'd married him for better or for worse."

Hal didn't smile. Hal was in a serious mood, and he continued. "Medical professionals can usually keep an emotional distance, but sometimes a death can really get to you. One of my worst calls was for three young fishermen—all brothers—who were drunk and going 160 kilometres an hour when they wrapped their car around a tree. One was dead, the others badly injured. It was night, and when I reached inside the car, brains covered my glove. Then one of the brothers grabbed my hand and wouldn't let go. That call still gives me nightmares. I don't know why ..."

I nodded. "I had a similar call. A woman came over the crest of a hill and ran into a semi. It shaved a foot off the side of the car and sliced her head in half. At the time I was working with Amber, a part-timer, and we spent an hour scraping up brains and other parts and putting them in a body bag. Ever since then Amber hasn't been able to eat beans—they remind her of that accident."

We stopped at a nearby café to buy two coffees and climbed back into the ambulance. We still didn't have a call, so we sat silently sipping our drinks. I pointed out the bumper sticker on a big SUV parked ahead. It said, "If you want to know whether there is life after death, just touch this truck."

Hal said, "Maybe we trivialize death as a way of avoiding dealing with it ..."

But before he could continue, the radio interrupted with a Code 3 for chest pain, and I forgot about our conversation for the rest of the shift. But that night I suddenly wondered how many dead people I'd seen, and that started a parade of memories and stories that I thought I had forgotten. They just kept coming and I couldn't stop them.

WORK EXPERIENCE

The call had come in at 7:00 a.m.—Code 3 for a possible Code 4. We were to back up an ALS car. We were just starting a ten-hour day shift, and the ALS crew was on overtime at the end of a fourteen-hour night shift. At the house a distraught woman led us to the kitchen, where a middle-aged man was lying sideways on two chairs. He must have died in the middle of the night as his body was in rigor mortis.

As I was examining the stiff body, the ALS crew arrived. I wasn't surprised to see them looking tired after their long shift, but I was surprised to see they were accompanied by a sixteen-year-old girl. She had long blond hair and was wearing an oversized ambulance jacket on top of a pink blouse, jeans and sneakers. One of the ALS guys,

Byron, introduced her: "This is Jody. Her school has arranged for her to ride along with us for work experience."

My partner and I nodded our greetings, and I said, "He's in rigor. You might as well clear." Jody didn't seem at all fazed by the sight of a dead man frozen in this strange horizontal position. In fact, she looked at him with interest.

Byron said, "Jody's handling everything really well. After all, this is her first time on an ambulance, and we've had four Code 4s this shift."

Four deaths in one shift! That doesn't happen very often, and the kid had probably never seen a dead body before. What bizarre luck. I wanted to ask her how she felt, but this was neither the time nor the place to start a conversation, so the ALS crew and Jody drove off, while my partner and I stayed behind to comfort the wife and arrange for the coroner.

To this day I find myself wondering whether that work experience changed Jody's life. Maybe it got her interested in medicine and now she is working as a paramedic, a doctor or a midwife. Or maybe she was traumatized and now needs therapy to overcome a deep-seated fear of death.

On the other hand, it probably didn't change her at all. From my experience with teenagers I'd guess that when she got home, her mother said, "How did it go, dear? Would you like to be a paramedic?"

And Jody replied, "Whatever. It's a weird job. They have to work all night." And then she grabbed some food from the refrigerator and ran off to her room to check her email.

STATUS UNCHANGED

We had been sent to the extended-care hospital for a routine transfer. We took the elevator up to the ward and were directed to an elderly woman's room. I knocked and opened the door, walked over to her bedside and asked her how she was feeling. No answer. I leaned closer in case she was deaf, but one good look told me she was more than deaf. She was dead.

I left my partner to start CPR and ran back to the nursing station where I had noticed a doctor. "Your patient has died!" I exclaimed. "You'd better bring the crash cart."

"I know," replied the doctor. "She's not to be revived. Now don't go making a fuss and disturbing the other residents."

"If she's dead and you don't want us to do anything, why did you call an ambulance?"

The doctor looked surprised. "Well, somebody has to take her to the morgue, and it might as well be you. If you'll just put an oxygen mask on her and tuck her in nicely, none of the other patients will even know she's passed away. Now please don't cause a commotion."

There was no point arguing. We put an oxygen mask on the body and covered it up warmly. As we wheeled the stretcher out of the ward, the other patients all looked up.

"Get better soon, Lucy!" they chorused. "Don't flirt with those cute ambulance men!"

One little lady looked up from her needlepoint and asked, "How is my friend doing?"

I wasn't sure what to say, so I answered, "Actually, her condition hasn't changed for a while now."

And I doubt it ever will.

THE ODESSA FILES

A policeman led us to the bathroom where a man was sitting on the toilet with his back against the raised seat cover and his pants down around his ankles. He looked like a healthy man in his mid-sixties, and to all appearances he was reading: his head leaned forward and his open eyes were gazing at the book he was holding on his lap. But as I got closer, I could see that his skin was grey, and when I lifted his head he stared blankly at the ceiling. I felt his neck for a carotid pulse, but there was nothing. He was dead.

But why? And why did he die there? I never saw the autopsy report, so I never found out. But for a few moments my partner and I stood there with the cop making guesses. He had obviously died suddenly. Perhaps he'd had a weak heart and the strain of having a bowel movement had triggered a massive heart attack. Or perhaps he'd read something so upsetting that his blood pressure rose high enough to burst an artery in his brain and give him a fatal stroke.

I lifted the book out of his hands and turned it over. He had been reading *The Odessa Files*, a thriller about a journalist who infiltrates a secret SS organization after the Second World War in search of a sadistic murderer. Could this story have reminded him of something that happened to his family? Had it made him really sad or angry? Had it killed him?

I still wonder about that man on the toilet. Was he the Nazi's last victim?

BREAKING UP IS HARD TO DO

Talking about toilets, I was told about a crew that was sent Code 3, possible Code 4 to an old apartment building. An ancient woman in a pink floral nightgown opened the door. "I think my Tom has finally died," she said. "I found him this morning in the toilet. It's a blessing, really—his heart's been giving him a terrible time for the last two years."

The old man was lying in a tiny room that held the toilet. It was obvious that he had died hours earlier as his body was now in rigor mortis, but in his last moments he had fallen to the floor and somehow wrapped himself around the toilet. When the paramedics tried to remove his body, they discovered that his arms and legs were locked around the bowl and the small space gave them no room to slide him back and out.

The driver said, "What do we do now? He won't loosen up for hours. Should we break the bowl?"

The attendant shook his head. "It's a perfectly good toilet—there's no point wrecking it. Just take his wife to the bedroom, close the door and start filling out the forms. Don't rush—ask lots of questions in a loud voice. I'll take care of things here."

The driver was confused, but as his partner was older and more experienced, he took the old lady's arm, walked her to the bedroom and closed the door behind them. He started to ask her questions. Behind him he could hear thudding noises.

After five minutes the bedroom door opened. The attendant smiled at the old lady. "Your husband's on the stretcher now, all nicely tucked in. Would you like to come and see him before we take him away?"

In the ambulance on the way to the hospital morgue, the driver asked, "How did you manage to get him out of there?"

"A little karate," the attendant replied. "The easiest way is to dislocate the joints."

Fortunately, bodies became the responsibility of the Coroner's Service soon after that incident, so if something like this happens nowadays, it's their problem.

LOOK WHO'S WATCHING

Brent and Steve were sent Code 3 to a car parked in front of a motel. They followed a police car into the parking lot, where a small group of people stood looking at a grey sedan. Its windows were covered with blood, and a ragged hole had been torn in the roof above the passenger's seat.

Brent said, "This looks terminal. I don't think we'll need a stretcher." He stopped the ambulance behind the sedan, and Steve grabbed the jump kit from the back.

The cop got out of his car at the same time and said, "I'll push the crowd back. You see if anyone is still alive."

They put on latex gloves, gingerly opened the driver's door and peered inside. It was a shocking sight. A man was sitting in the passenger's seat with a shotgun between his knees. He had put the gun's muzzle in his mouth and pulled the trigger, blowing off his face and the top of his head. Now bits of bone, skin and brain dripped from the roof into the gaping wound.

The paramedics stared at the horrifying scene. Then Steve said, "I think we're being watched."

Brent was confused. "What? Who?"

Steve nudged him. "Look down there in front of the driver's seat."

Brent looked down. An eyeball was sitting on the mat, its blue iris staring upwards.

"See?" Steve said. "It's watching us!"

Brent was so tense he couldn't help himself. He started to giggle, and in a moment both he and Steve were helplessly shaking with laughter. They hung on to each other for support and bit their lips in a desperate effort to get a grip. After a few minutes their laughter subsided, and they were able to close the door and walk over to the cop.

"It looks like a suicide all right," Steve said. "One male victim with a gunshot wound to the head. No one else is in the car. You won't be needing us, so we'll clear and get on our way."

The cop noticed that the paramedics had tears in their eyes. "Pretty sad, eh?" he said. "You never really get used to these things, do you?"

THE CORPSE THAT WOULDN'T STOP TALKING

An experienced Advanced Life Support crew was sent on a Code 3 for a possible Code 4, backed up by a Basic Life Support car staffed by two younger paramedics, Fred and Sarah. When they arrived at the scene, they were led to a bedroom where an older man was lying staring at the ceiling. He was not breathing and no pulse could be found. The young paramedics started CPR while the ALS crew hooked up their monitor. A flat line appeared on the screen.

"Asystole," said Vince, who as the senior ALS attendant was in charge of the call. "Let's start a line and give him some epi." For the next fifteen minutes they ran through their protocol, intubating the patient and infusing more epinephrine and atropine. At first they managed to get the heart to contract a few times, but it would not revive and the screen again showed a flat line. Finally Vince said, "There's nothing more we can do. Cease compressions. I'll go and talk with the family."

The BLS paramedics lifted the body onto their stretcher, covered it with a blanket and cleaned up the bedroom. They gave their condolences to the patient's wife and sons, took the stretcher outside and loaded it into their vehicle. Vince followed them outside, winking at his partner, then spoke to the younger crew. "Take the body to VGH emerg. Once you're there, you can get it admitted to the morgue. We'll follow you."

Fred and Sarah shut the back door of their ambulance, got in front and started off. For a few minutes the trip was uneventful. But then they heard a low moaning coming from the back.

"Ooow, ooow, ooow."

Fred was startled. "What's that?" he said. "He can't still be alive?"

"I don't know," Sarah replied nervously. "We'd better check." She turned on the emergency lights and pulled over to the curb. The ALS car following also pulled to the side and parked.

Fred climbed into the back, uncovered the man's face and checked for breathing. No sounds. He then felt for a carotid pulse. No heartbeat.

The ALS crew peered in the side door, and Vince asked, "What are you doing? He's dead. No need for a second opinion."

"Sorry," Fred said. "I thought I heard noises."

"That's okay," Vince reassured him. "We all make mistakes. Let's go. We don't have all night."

They got back in their vehicles and drove off. All was quiet for two minutes, and then noises started coming from the stretcher again, but this time they were louder: *"OOOW, OOOW, OOOW!"*

Fred's hair almost stood on end. "Did you hear that?"

"I heard it. He must be alive!" Sarah pulled to the curb quickly and hit the brakes. Once more the ALS car parked behind them.

Fred put his stethoscope on the man's chest and listened for heart and respiratory sounds. Absolutely nothing! Again the ALS crew opened the side door and Vince said, "What on earth are you doing? Let it go—the guy is gone."

"No," Fred protested. "He can't be. Both Sarah and I heard him moaning."

"Look," Vince said, "we'll show you. We'll get the monitor and you can see for yourself." In a few moments the leads from the cardiac monitor were reconnected to the patches on the patient's chest. "See? No activity. Just like I said, he's dead, dead, dead. Now stop fooling around. Let's go!"

Once more the two cars headed toward the hospital, Fred watching the body nervously in the rear-view mirror. But it lay perfectly still and all was quiet—for a few minutes. Then the moaning began again, this time in a singsong voice: *"OOOeee, EEEooo, wowowow, EEEOOOEEE!"* Fred broke into a cold sweat and Sarah's hands trembled on the wheel. She braked sharply and once more Fred got in the back and uncovered the patient. Still no respiratory or heart sounds!

Fred was almost in tears. "It's really strange. I tell you, he was moaning but there are no signs of life."

But this time the ALS crew were standing in the doorway, grinning from ear to ear. "Let the experts have a look," Vince said. He and his partner got in the car and lifted the patient. Underneath the body was a black two-way radio. Laughing, Vince picked it up and clipped it to his belt. "You have to learn to trust us," he said. "We told you he was dead!"

Two crews were sitting in the station waiting for calls. Cathy asked Marv if he was in a relationship.

"Nah, it's just me and my two trucks."

Cathy said, "You'd be popular on Salt Spring Island. There are two women for each man and by spring they're running out of firewood. Just put out the word and they'll be fighting over you. All you'll have to do is split a few cords and then you can warm up the bed."

Marv laughed and said, "By then I'll be too tired to do anything more than lie still and groan."

Cathy shrugged, "That's about all that the men I go out with do anyway."

Paramedics almost always respond to their patients with marvellous sensitivity and compassion, but occasionally something happens that results in a lapse in professionalism. Something that usually has to do with sex.

ALL THE WRONG VIBRATIONS

A crew was called to an apartment late at night for an unspecified problem. A young woman wearing nothing more than a sheet met them at the door and without a word led them to her bedroom, climbed onto her bed and lay down.

The attendant was a bit confused by her obvious good health. "What seems to be the problem?"

"I don't know how to tell you this ..." There was a long, embarrassed pause. "I've got my vibrator stuck inside."

The paramedic stifled a laugh. "Well, why don't we just take you to a hospital and let a doctor get it out for you?"

"You don't understand. The damn thing is still turned on, and I can't stand it anymore."

"So what would you like us to do?"

"Could one of you just reach inside and either get it out or turn it off?"

The attendant was a little dubious; this procedure wasn't taught in any course he'd taken. But he put on a medical glove, reached in and tried to extract the vibrator. Unfortunately it was wedged in too tightly to be easily dislodged. Next he felt for a switch, but it was out of reach. Finally he managed to unscrew the base, and as the batteries fell out, the vibrations stopped, and the woman groaned in relief.

The attendant looked at her with a big grin. "I know your problem."

"What problem?" She was feeling too humiliated to listen to a lecture.

"You've got the wrong batteries in that thing," he said. "The ones you're using are the kind that just keep going and going and going!"

The Helpful Patient

Because the ER was pretty quiet, most of the nurses had left for supper. I had just given my report and was washing my hands when I heard Nicole call across the nursing station.

"Anna, it's time to check everyone. I'll do all the patients on this side."

Anna called back, "Okay. It looks like there are only men on my side. I can do the men."

Just then an emaciated old woman sat up in one of the monitor beds, ECG leads dangling from under her white gown. "Wait! Let me do the men!" she cried. "I'll do all the men!"

ALL STUFFED UP

The woman was suffering from lower abdominal pain, but she was reluctant to explain what had happened, even to the female paramedic. Finally, in the back of the ambulance, she blurted out the story.

It was their wedding anniversary, and she and her husband had decided to be adventurous. She had let him stuff her full of fruit, but now they couldn't get them out again and she didn't know what to do.

The paramedic sat beside her, dutifully filling out a report while pinching herself to keep from laughing. She was going to say that next time it might work better if the whipping cream was applied first, but she decided that perhaps it was better to say nothing …

FULL-SERVICE DISPATCHING

The man on the other end of the phone sounded desperate. "You have to help me! I'm going crazy!"

"What is the problem, sir?" The female dispatcher had her pen ready.

"I need some relief really bad. If I don't get some relief, I don't know what I might do to myself. You have to help me!"

"Don't worry, sir. I'm here to help you, but you have to tell me what your problem is and give me your address."

"No, I can't tell you my address. But don't go away. I need to talk to somebody and you sound so nice. You won't stop talking to me, will you?" The man's voice was rising and he sounded increasingly agitated and short of breath.

"Of course I won't go away. But you must calm down. Now please tell me where you are so I can send an ambulance."

"I don't need an ambulance! I just need some relief. And I love your voice. You're just perfect!" The man was gasping now.

All of a sudden the dispatcher grasped what was going on. She began laughing so hard she could no longer talk and had to hang onto the desk to keep herself from falling out of her chair.

"Oh, yes, yes, yes!" he cried out. "That was wonderful! What a relief! How can I ever thank you? I'm so glad I called." And with a last ecstatic sigh, the caller hung up.

When the dispatcher recovered, she thought about the call. When someone calls for an ambulance, the procedure is usually quite straightforward: if the ambulance takes them somewhere, they get charged for the service. But if the caller comes on his own, how can you charge him?

Emergency Rooms

We had just brought a patient into the emergency room and I was giving my report to a tired nurse when a confused patient in an adjoining cubicle climbed off his bed, staggered over to the nursing station and ordered a beer. He had untied his gown and it fell open, exposing his privates. The nurse groaned, saying, "Oh, gawd!" and covered him up. Then she tied his gown, took him by the shoulders and gently but firmly steered him back to bed.

I have tremendous respect for emergency room staff, whose work requires a rare combination of intelligence, technical skills, dedication and compassion. Not everyone can perform a complex medical procedure one minute, comfort a dying patient the next, and then move on to another bed to wash urine and feces off an old drunk.

NO ROOM FOR ERROR

One evening Mark and I had just lifted a patient from our stretcher onto an ER gurney when a nurse called for help. A patient's heart had just stopped. We immediately started CPR, and we were joined by a doctor and two more nurses who began to intubate the patient, analyze the ECG data and infuse drugs.

A few minutes later a monitor alarm sounded from the adjoining bed. Another patient was in arrest! Another doctor and two more nurses appeared and started arrest protocols. Then an alarm went off on a third bed. Three simultaneous arrests!

Now every nurse and doctor in the ER was involved, and an ambulance crew that had been ready to clear the hospital was called back to help. Lab technicians came, drew blood samples and left. Although fifteen people were working on the patients, there was no chatter. Everyone knew their role; reports were given and instructions issued in calm, quiet voices.

As soon as the first patient revived and was stabilized, she was whisked away to the critical-care unit. Then a second regained a pulse and was also taken to the CCU. The third took longer, only maintaining a pulse with the help of an external pacemaker. But once the pulse was strong enough, he was sent up to the operating room for heart surgery.

It's a great feeling—and a great privilege—to work with such professional, caring people!

JUST PULLING YOUR LEG

Every once in a while, though, something happens in ER that cracks everybody up.

One morning Mark and I were making up our stretcher in the hallway next to the ER waiting room. Across from us a patient with a bloody bandage on his head was lying on another ambulance stretcher. A double amputee, he had flipped his wheelchair while trying to get over a curb. The real damage, however, wasn't to his head—that was only a minor laceration—but to one of his artificial legs. Although a blanket covered most of his body, the leg jutted out at a strange angle. It was badly broken.

After ensuring that his patient was comfortable, the attendant had left him and walked over to the triage desk to give his report. Behind him, while keeping one hand on the stretcher, his partner had turned his back to the patient and was chatting with another crew.

At this moment a senior orthopaedic surgeon walked past me on his rounds, trailed by a gaggle of interns. Each of them was holding a pen and notepad, ready to jot down the expert's wise insights.

"What's this?" The surgeon had spotted the broken leg. "How unprofessional! Emergency personnel should apply immediate traction to a broken leg and then splint it."

He strode over to the stretcher, placed a foot against the undercarriage, gripped the leg below the break, and pulled. The stretcher lurched forward, the leg flew off and the surgeon flipped onto his backside.

The interns gasped and ran for cover. There they covered their mouths and laughed until they cried.

No Knobs

Although emergency rooms are usually serious places filled with sick patients and busy staff, the nurses working in emergency were always cracking me up. Once when I was filling out my paperwork at the nursing station, I overheard Jo say to Jessica, "No knobs. Didn't I tell you I don't do knobs? Get John to stick a catheter up his bladder."

And Jessica said, "Which one is John? Is he the orderly or the disorderly?"

UP YOURS

My partner and I were cleaning our stretcher in the hallway when a nurse rushed out of the ER. As soon as the doors closed behind him, he burst out laughing. A minute later Marie, the triage nurse, came giggling back to her station and waved us over to tell us what was going on. They had just admitted a university student who was complaining of rectal pain. He had gotten drunk at a party the night before, passed out and woke up feeling like he had a stick up his bum.

The student was put on a bed in a cubicle and told to turn over on his side and bend his knees so a doctor could examine him. Dr. Robins ducked behind the curtains and greeted the patient. He then put on a glove, lubricated it and inserted his index finger. Right away he felt something hard, possibly metal or glass. Very strange! He got a speculum and opened the patient's anus. Then he leaned down to take a close look and almost fell over with surprise. A bright light was shining back at him!

The doctor smothered a laugh. It appeared that the student's "friends" had stuffed a flashlight up his bum while he was sleeping! This was too good to be missed. He told the patient not to move; he would need a second opinion on the best way to extricate the foreign object. Then he leaned around the cubicle curtains, put his figure to his lips and one by one waved over the entire emergency room staff.

Marie chuckled. "Now I'm a believer. Sometimes there is light at the end of the tunnel!"

MALPRACTICING MEDICINE

Our patient was an old woman with dementia who had fallen and lacerated her scalp. Marv and I took her to a room at the back of emergency where a nurse, Rita, held her head while a young doctor quickly stapled the wounds. He had no sooner left than Rita pulled us to one side and whispered, "What an amateur! Look at the mess he made. He put a staple in the wrong place and missed two cuts entirely. We could do much better."

"Why don't we?" Marv said.

"Because it could cost us our jobs, that's why," Rita said. Then she paused to consider the idea. "Well, okay, but we'd better watch the door ..." And she picked up the skin staple remover. Just then the doctor re-entered the room, and she whipped the staple remover behind her back.

The doctor didn't notice. "Mrs. Thompson hasn't had a tetanus shot in fifty years. We should give her one."

"No problem," Rita said. "That will only take me a jiffy."

After the doctor had gone again, she gave the patient a tetanus shot. Then I guarded the door while she and Marv restapled the lacerations and bandaged the woman's head and jaw with surgical crepe. "Hey, this is easy!" Marv said.

"While we're at it," Rita said, "why don't we staple the ends of the bandage together?"

"Sure." Marv pressed the trigger again. There was a dull *thunk*, and the old lady blinked. Marv jerked his hand back, alarmed. "Oh-oh! I've stapled the bandage to her skull!"

Rita collapsed against a cabinet. "If you want me to keep quiet about this caper, you'll have to buy me free coffee and flowers for a year!" After Marv promised to do anything she wanted, Rita helped him pull out the staple and refasten the bandage with tape. All this time our victim had sat perfectly still, apparently happy with the unusual attention.

We hoped Rita would keep her end of the bargain and that would be the end of the story. But by the time we brought in our next patient, a big sign had been put on display at the nursing station: TO ERR IS HUMAN, TO STAPLE DIVINE. And when we wheeled our stretcher through emergency, all the nurses grinned and tapped their heads.

Medevacs by Air

The British Columbia Ambulance Service provides emergency medical services to more than 4.5 million people scattered over an area three times the size of Italy. It is an area that includes some of the most spectacularly diverse geography on earth, and to operate here in all conditions including blizzards, forest fires and floods, BCAS not only maintains a fleet of over five hundred ambulances and support vehicles, but also owns or hires a small fleet of helicopters, planes and boats.

I love any kind of flying, and on my days off, I would often offer to work extra shifts with the air ambulance. My favourite planes were the Cessna Citation business-class jets that were the backbone of the BCAS fleet because they were not only much quieter than helicopters and turboprops but also fast and acrobatic. The pilots were top-notch. Many had given up better-paying jobs with commercial airlines to work for the Ambulance Service because it gave them the opportunity to fly these cool machines over really beautiful country.

I have many memories of those flights: skimming over glistening white glaciers before spiralling steeply down into the shadows of the Bella Coola Valley; cruising close to Mount Robson, its sheer cliffs red in the sunset; slipping over the green, mist-shrouded islands of the northern coast, untouched and unchanged for thousands of years; flying high over Castlegar, where a sea of white scalloped clouds extended from horizon to horizon, only interrupted by the golden spires of mountaintops; and flying into Vancouver from the north at night and watching the blackness suddenly peel back to reveal the city shining like a million sparkling jewels.

On one flight we took off empty from Vancouver as we were heading up to Haida Gwaii to pick up an injured fisherman. The pilots asked my partner and me, "Do you mind if we have some fun?"

Did we mind? "Go for it!" we said enthusiastically, and the jet took off from the airport, climbed sharply, and then stood on its wing tips and banked ninety degrees like a fighter plane. We raced low over the Salish Sea, the plane dipping and weaving between the sun-dappled islands.

ONLY HALFWAY TO HEAVEN

It had been a lovely winter day. On the flight from Vancouver to northern BC, the setting sun had bathed the ocean, islands and mountains in delicate pinks and golds. Now we were flying back south in the darkness. The lights were turned down in the jet's small cabin, and it felt very peaceful; in the cockpit the pilots talked quietly in front of the glowing control panel. Dan was sitting beside the stretcher, holding the hand of a terminal cancer patient and reassuring her as she talked about her fear of dying.

I sat at the back of the cabin, staring out the window, acutely aware of how the moment was simultaneously sweet and sad, and thinking of my own situation and a relationship that wasn't to be. Then I wrote this:

There are stars above and stars below
On the unseen earth a few lights glow
The world is upside down tonight
For I'm caught up in an endless flight.

Stuck in a capsule suspended in space
Gambling with gods on a crazy race
Up in the air, though I don't want to fly
Wishing I was with you, though you're saying goodbye.

Pale clouds float on a pitch-black sea
A red streak remains where the sun used to be
My life rushes on through beauty and pain
If I open my heart, love may keep me sane.

The wing tips are steady, outlined with snow
I'm strong deep inside, where aloneness must grow
But I'm still sad to see your warm brightness go
Even with soft stars above, and stars shining below.

FLYING WITH THE INFANT TRANSPORT TEAM

Occasionally I worked with the Infant Transport Team (ITT), which is based at BC Children's Hospital in Vancouver and specializes in transporting high-risk obstetric patients, premature babies and very sick children. I was always impressed by the quality of the people who work on these teams; they bring an amazing amount of love, care and skill to the job. They sometimes spend hours beforehand making sure that premature babies are stable enough to ensure a safe transfer. My job as an intermediate-level paramedic was to drive the ambulance, assist the specialist and check and ready the equipment, including the heavy transport incubator with its Plexiglas cover, heater and oxygen supply.

As a bonus, the Infant Transport Teams also spend half their time flying around the province bringing high-risk patients from remote towns to hospitals in Vancouver and Victoria. So I was thrilled to get a call from dispatch one morning asking me whether I wanted to do a shift with them.

"Of course!" I said. "Will we be going flying?"

The dispatcher laughed. "Only if you trip."

TESTING, 1, 2, 3 ...

Sometimes we didn't take planes to patients; instead, planes brought the patients to us. On one occasion we were sent to the float-plane dock in the harbour to pick up a middle-aged man who was having trouble breathing and seeing. The pilot told us that his passenger had hired the plane to fly him to a fishing camp on a remote lake. This was his first week-long solo trip in the bush, and he was very nervous about running into a grizzly. So before setting off, he had bought a big can of bear spray.

The pilot unloaded the fisherman and his gear at the lake and was just about to take off when he saw that the man was now rolling about on the ground. He had decided to test the bear spray on himself to see whether it worked.

It sure did!

TRUMP'S HELICOPTER

When I was nineteen, on holiday in Trinidad, I learned that you aren't really safe anywhere. Never.

I had wanted to explore the island's Atlantic coast and was directed down a path to the shore. When I walked out of the bush, I found myself standing all alone on a gorgeous beach. The blue sea stretched to the horizon, a gentle white surf was breaking against the golden sand, and above me tall coconut palms swayed in the warm breeze. It was paradise!

It was shady under the trees, so I lay down and closed my eyes. Everything was perfect: the smell of tropical flowers and salt air, and the sound of falling surf mixed with the cheerful chatter of birds. So sweet, so relaxing ... *thump!* A huge coconut landed right beside my head! Terrified, I leaped up and ran to the middle of the beach. That coconut had almost killed me!

Back in Port of Spain, I told my new Trinidadian friends what had happened. The girls cracked up. "Fuh true? Man, you Canadians don't know anything! Before you walk under a coconut tree, you look up. If the coconuts are bright green, they are still growing, so you're safe. But if the coconuts are brown and ripe, you stay away!" From the way they looked at each other and smiled, I knew two more beautiful women had just written me off as an idiot, so I made my excuses and slunk away.

You would think that I would pay more attention to safety after this humiliating experience. But I didn't.

I was working on an ALS car with Tom when we were sent Code 3 to the airport. We were to take a helicopter to a back road for a motorcyclist who had run into a logging truck. After a security guard waved us through the gate, we unloaded our gear outside a private hangar belonging to a company under contract to the BC Ambulance Service. A pilot led us past busy mechanics to a big, gleaming Sikorsky helicopter parked out back on the tarmac. "You're in luck," he said. "We just bought this fancy machine and you'll be our first passengers."

I was suitably impressed. "Wow, it looks brand new!"

"Actually, it's seven years old, but we just gave it a new paint job. The last owner was Ivana Trump."

By now we had reached the helicopter, and the pilot opened the side door. "As this is your first time aboard, I'd better explain the safety features." He started talking, but Tom and I weren't paying attention. Instead we were gawking at the luxurious cabin. Although two seats had been removed and replaced with a stretcher platform, two large executive chairs remained, each covered with buttery-soft leather. The armrests and the walls were trimmed in teak, and the sophisticated electronic console that ran down the centre of the roof looked capable of picking up any TV or radio channel on earth.

The pilot saw me staring at the console. "Don't worry about all this stuff," he said. "All you need to know is that you can talk to my co-pilot and me right here on Channel 1 and to your provincial dispatch on Channel 5. Okay, let's go—it sounds like that motorcyclist is in pretty rough shape."

We secured our equipment and settled into the leather armchairs. The pilot shut the door beside us before joining his co-pilot in the cockpit and starting the engines. In a few minutes we were airborne, quickly rising over the Salish Sea on our way to the middle of Vancouver Island.

I grinned at Tom. "I can't believe that some of the richest asses on the planet have sat in this seat!"

"No kidding," Tom replied. "This is more a heavencopter than a helicopter."

Our reverie was interrupted by the pilot's tense voice. "I hate to tell you this, fellas, but we're losing hydraulic oil fast. I'll try to get back to land, but we may have to ditch in the salt chuck. Get ready for a water landing."

Tom and I looked at each other in horror. Where were the life jackets? Would the door open underwater? How would we unlock it? Frantically, we started to check out the cabin, looking at the doors and windows and feeling under the seats. But we had no time to get oriented. The helicopter was descending quickly, and before I had pulled out my life jacket, the pilot had landed back on the pad.

That little adventure finally taught me to pay attention to safety! I'm sure you've taken commercial flights where half the passengers can't be bothered watching the flight attendants point out the escape routes and demonstrate how to blow an emergency whistle. Not me. I listen to every word they say and carefully count the number of rows to the nearest door. I know airplane travel is very safe—much safer than going by car—but you never know. With my luck a coconut could fall into one of the engines ...

I'll bet you want to know what happened to that poor biker, but I can't tell you because I didn't ask. I have a motorcycle licence myself, and I have to admit that few things beat riding along a winding mountain road on a warm summer day. But gravel roads are tricky, and it only takes one blind corner and a logging truck to turn a motorcycle into a murdercycle. Still you never know—he may have survived. An ambulance might have reached him in time, and he might have made a full recovery. I hope so.

DOUBLE RAINBOWS

My memories of helicopter travel include the surreal feeling of landing on the roof of Vancouver General Hospital in the middle of the night with a critically injured trauma patient. After we rushed him downstairs, I returned to the waiting helicopter on the roof, a strange space suspended above a sleeping city, a world of shadows and silence after the bright lights and bustle of the operating theatre.

Another time, we were asked to help a crew unload an air ambulance patient at the Victoria General heliport. There, the pad was out back of the hospital, surrounded by grass sprinkled with rotund rabbits and, farther back, by tall trees. It had been raining that afternoon, and as we waited for the helicopter, the clouds lifted, the sun came out and two rainbows appeared, one above the other. The scene was astonishingly beautiful: coloured bands rising up from the green woods to form spectacular arches over the landing zone.

I jumped out of our ambulance and dashed for the hospital. Inside, I ran from office to office begging for a camera. But this was before the days of smartphones, and there were none to be found. So I ran back outside just in time to see the helicopter descend. It was painted in BCAS colours, white with a red stripe, and it came down slowly right in the middle of the double rainbows. Pure magic.

But I almost cried. It was the perfect photo and I didn't have a camera!

Medevacs by Sea

The Ambulance Service also has to be able to reach anyone who is hurt or sick on one of the forty thousand islands that dot the rugged BC coast. Helicopters, police boats and the Coast Guard's hovercraft are used to access the remote islands, and private water taxis are contracted to service the larger Gulf Islands. Typically the water taxi companies reconfigure their boats to allow a portable stretcher to be fastened down on some of the seats. Whenever a water taxi is needed, an ambulance dispatcher pages the captain, who picks up the ambulance crew and their equipment from a wharf and takes them to their destination.

On a nice day these trips can be quite pleasant with gorgeous scenery and the frequent sight of cormorants, seals and families of playful otters. At night it becomes a different world; dark rocks can blend with a black sea, and it takes a skilled pilot to navigate in the dangerous shallows and treacherous currents. Sometimes the weather is rough, and the unfortunate patient is jolted up and down with each wave. This can be misery for the patient, but it's also no fun for the attending paramedics, who are anxious to get their ward out of the boat and into the relative warmth and safety of an ambulance.

Most trips are uneventful. But this is British Columbia ...

GOOD ENERGY

My part-timer partner Sally and I took a water taxi to one of the outer Gulf Islands to pick up the local doctor, who had been picking berries when a small thorn became stuck in his eye. Despite his best efforts, it had lodged out of reach at the back of his eyeball, and by the time we arrived, he was in considerable discomfort. Our job was to transport him to Victoria General to get the thorn removed.

The doctor and his wife were an attractive couple in their mid-thirties, and he looked really healthy even with both eyes covered by a cloth bandage. We put him on our stretcher, lifted him into the boat and fastened the stretcher down. His wife sat next to his head.

I asked him, "Could we give you some Entonox for the pain?"

The doctor shook his head slowly, and his wife said, "He's already taken some painkillers. They're helping a bit, but the itchy, scratchy feeling is driving him crazy. But maybe I can help him relax." She turned away from us, leaned toward her husband and held her hands palms down over his eyes.

I teach reiki, a form of energy healing, so I knew she was giving him an energy treatment. Most scientists doubt whether these methods are any better than placebos, but so what? Placebos work half the time, and like marriages or your relationship with your kids, they work best when you believe in them. Reiki works for me, so I'm happy to practise it. In fact, therapeutic touch, another form of energy work, is taught in many nursing colleges because it often helps to reduce patients' pain and fear. However, because I wasn't licensed to use it as a paramedic, I only treated people in my spare time.

But this time I made an exception. After all, the doctor was the medical authority on board, and if he approved, who could object? So I explained that I was a reiki practitioner and asked whether I could help.

"Yes, please," his wife said. "You can take the foot end."

At that point Sally jumped in. She's a clairvoyant, which is not so unusual in these parts. "I'll steady and strengthen the energy flow." She lifted her arms.

All this time the captain had been standing in the bow cockpit steering the boat, but as he was only a few metres from us, I'm sure he had heard our voices through the thin curtain separating us. Now he pulled back the curtain and looked at us in alarm. I'm sure he thought he was transporting a group of witches. I could see he was about to say something, but he checked himself and turned back to his work, and the rest of the journey passed in peaceful silence.

BAD ENERGY

Another time we took a water taxi to Lasqueti, an island some eighty kilometres northwest of Vancouver. As we approached the shore, a camouflaged speedboat raced out and pulled up next to us. The long-haired young guy behind the wheel shouted to our captain, "Can you turn off your radar? It's creating too much negative energy. The bad vibes are ruining lunch for half a kilometre in every direction."

The captain was furious. "Bugger off, you little creep! Get out of here before I beat you to a pulp with my fish club."

The hippie shook his head and gunned his motor. In a few minutes his speedboat had disappeared around a headland.

I was surprised at the exchange. "What was that about? I've never heard of radar affecting people's appetites."

The captain snorted. "It's got nothing to do with bad vibes or lunch. They grow acres of weed on Lasqueti, and I bet they have radar detectors set up to watch out for the Mounties. He's pissed off because I'm triggering all their alarms!"

Off the Record

We were working an afternoon shift on my mother's birthday, and because our station covered the area my parents lived in, I asked the dispatcher whether my partner Mark and I could drop by and give her our best wishes.

"No problem," he replied. "I've got nothing on the board right now. Have a good visit."

"Thanks," I said. "If you need us, we'll be available on our portable radio."

It was a sunny June day, and soon my mother, father, Mark and I were sitting around their patio table eating birthday cake and drinking tea. We'd placed the black portable radio in the middle of the table, but it was absolutely silent.

I turned to Mark. "It's unusually quiet today."

"Let's hope this lasts," he said.

The weather was perfect, my parents' garden was beautiful, and we were having a lovely time. As we chatted, the minutes turned to hours. Finally I looked at my watch. "Good Lord, it's almost five o'clock! Maybe it's time to go."

Back in our ambulance, I picked up the microphone and went 10-8. The dispatcher didn't sound pleased. "Sixteen Bravo, what happened to you? I've been calling you for hours!"

Oh-oh! I quickly checked the portable radio. "My apologies! I had the portable set on the wrong channel. Just give us all the calls for the rest of our shift." And he did. We worked straight through to midnight.

That was embarrassing. Still, it is one of my favourite on-the-job memories. My mother died a year later.

Part v: Down and Out

Compensation Claims

I'm very grateful that injured workers are given financial and medical assistance in British Columbia. In fact, my life would have been very difficult without this help. Nevertheless, it's not easy to collect compensation because the administrators are (rightfully) paid to be suspicious.

THE COLD CREAM ACCIDENT

The elderly couple had just finished shopping and were driving home when the husband had a heart attack and lost control. The car hit the ditch and flipped over, and when we arrived at the scene both husband and wife were hanging upside down, held in place by their seatbelts. What made this situation truly bizarre was that among their purchases was a large jar of cold cream, and it had broken in the crash. Now they were covered with white, slippery cream.

I got on my knees and pried open the driver's door. Holding up the greasy husband with one hand, I cut open his seatbelt with my scissors. He fell heavily into my arms and I slid him onto a backboard so my partner could give him oxygen and haul him over to the ambulance for treatment. Then I went to the other side and repeated the process with his wife.

The next day I could barely get out of bed. Not only were the muscles strained in my arms and back, but my testicles had swollen to the size of a grapefruit. I booked off work and saw a doctor, who recommended cold packs and a week of rest.

The Ambulance Service had no objection to me staying home, but I had to file a Workers' Compensation Board claim in order to get paid during my time off work. I duly filled out a claim and sent it in. A couple of weeks later, my claim was rejected because the board's medical advisor didn't think it was possible for me to be injured in this way while kneeling.

"So what's their explanation?" I asked my partner. "Do they think you kicked me in the crotch or that I hit myself with a hammer?" I collected witness statements and filed an appeal, but it was three months before an (expensive) appeal board tribunal finally ruled in my favour. At the time the incident was more amusing than serious as the injury healed and never troubled me again. The real problems started a few years later.

DOWN FOR THE COUNT

A few years later Susan and I were injured on a routine transfer. It was a very windy day, and just as we began lifting our loaded stretcher into the ambulance, one of the back doors blew across the head of the stretcher, almost hitting our patient. Our lift stopped with a sudden jerk, wrenching our backs. We were both sent for physiotherapy, and while Susan was able to return to work after four months, it took me seven months to recover. During most of this time I couldn't walk half a block without sitting down, and I remember leaning against a building and wondering whether I would ever be able to work again.

That injury permanently weakened my back, and I knew that it would only be a matter of time before a heavy lift or another accident damaged it again. Perhaps I should have resigned from the Ambulance Service right then, but at the time I couldn't afford to stop working and retrain for another career. Besides, where would I find another job that was half as interesting?

Over the years my back was injured more and more frequently, until finally I suffered my fifteenth back injury. It was nothing dramatic, just another heavy lift, but I was never able to return to work.

The management and the union generously offered to retrain me to be an ambulance dispatcher, but I declined. Not only did my back hurt too much to sit still for long periods, but I didn't think I was cut out for the job—ambulance dispatching requires excellent multitasking skills, and I have trouble cooking a three-course meal without burning something. So I retired from the British Columbia Ambulance Service and went back to school.

Was It Worth It?

I could sum up my ambulance career as twenty-one years and twenty thousand calls. It was also a fascinating and exciting roller-coaster ride that showed me the best as well as the worst sides of humanity.

Ambulances are rarely called for minor problems; when paramedics arrive, the people waiting for them are usually very sick or badly injured, frequently frightened and in intense pain, and often dying. Although paramedics learn to approach their jobs with professional detachment, some incidents are so shocking that the memories never go away.

I was lucky because I didn't have that many horrible experiences, and I don't suffer from post-traumatic stress disorder (PTSD). All the same, the job gradually wore me down, and my confidence weakened with every back injury. Towards the end of my career I didn't know whether I would make it through the next shift or fall down a flight of stairs on top of a patient. By then I no longer enjoyed going to work—the anticipation had been replaced by anxiety—I had burnt out.

Was the experience worth all the injuries and stress? They say a good education is expensive, so I can't complain. Ambulance work taught me so much; perhaps I even gained some wisdom.

Judgement Day

I didn't save a lot of lives because I wasn't an Advanced Life Support paramedic dealing with critical heart conditions. But as intermediate-level paramedics, my partners and I responded to a wide range of problems including traumas, diseases, mental illnesses and overdoses. We helped and reassured thousands of people who were sick, scared and confused, and I believe that along with doing useful work, we did a good job. Although occasionally, I had doubts ...

After I left the British Columbia Ambulance Service, I would sometimes lie in bed at night thinking about what would happen when I died. I thought of the people who have had near-death experiences and had found themselves going through a dark tunnel toward a bright light. They arrived at a place of indescribable peace and beauty to find their parents and friends waiting for them. I imagined rushing eagerly down a tunnel toward that light, anticipating a joyous reunion with my family, only to discover that the light was coming from a white neon sign above a frosted-glass door.

At first I think it's an immigration office, but then I see the door is labelled OFFICE OF THE FINAL JUDGEMENT. I stop in surprise, but because the sign is flashing ENTER HERE, I step gingerly inside. It doesn't look at all like an office—it's just an empty space surrounded by shifting clouds. A grim-looking, white-haired fellow in a long robe is staring intently at a holographic tablet, and for a few minutes both of us are silent. But then his steel-grey eyes focus on me.

"Ah, Graeme Taylor," he says. "We've been expecting you. I'm a recording angel, here to pass judgement on your life."

"You judge people?"

"Of course we do. You've known that from the time you were three and your mother told you about heaven and hell. She also taught you right from wrong, so you have no excuses."

"I always tried to do my best," I say nervously.

"In that case, you shouldn't have anything to worry about."

He pauses to look at the stream of images and charts dancing above his tablet. "I see you were pretty normal as a kid, but then you started to party and dropped out of university. In general you wasted your twenties, though later in life you did something useful, so your youth and your old age pretty well cancel each other out. However, I am more interested in your career as a paramedic. During that time you had a lot of responsibility, so everything you did really mattered. You'll be interested to hear your final scorecard: on one hand, you saved or helped to save seventy-nine lives, which is very good."

"Ah, yes," I add modestly, "when I worked in the Interior, I had to deal with a lot of bad trauma, and I also worked on busy cars in Vancouver and Victoria. Sometimes we would have two or three cardiac arrests in a single shift. Of course, I can't claim all the credit—I worked with a lot of first-rate paramedics, nurses and doctors ..."

The angel doesn't seem impressed. Instead he looks at me sternly and says, "But on the other hand, you killed or helped kill eighty-two people through negligence and malpractice. I'm sorry, but you get a failing grade. I'll have to send you downstairs."

"That's not possible!" I protest. "Sure, a lot of our patients died, but they were old or sick or had been run over by trucks. Paramedics can't save everybody."

"All right, just so you can see I'm being fair, pick any number from one to eighty-two, and we'll see what happened to that patient."

I'm starting to panic. "Uh ... uh ... number thirty-nine."

"Okay, look at the tablet—I'll replay the whole ambulance call for case number thirty-nine from start to finish. By now you've been working as a paramedic for thirteen years. It is 10:56 in the morning when the hotline rings at the station. Your partner answers and it is an emergency call for a Code 3, possible Code 4. He tells you that it's an emergency, but you are on the toilet in the middle of a dump, as you say. However, rather than hurrying up you take your time, with the result that you arrive forty-two seconds too late to revive your patient."

I stammer, "But I was beginning to get hemorrhoids, and I was afraid I would start bleeding if I tried to rush. I didn't want to go on a call with a wet stain on the back of my pants."

"So which was more important, your pride or his life?"

"Uh ... his life, I guess."

The angel has the same prim expression my high school math teacher used to wear when I had the wrong answer ... again. "Exactly!" he says. "A clear case of criminal negligence."

"But I was always kind to my patients. Doesn't that count?"

"Not good enough. You were paid to practise medicine, not to chat about the weather. And you weren't always nice. It says here you were a macho paramedic who enjoyed making tasteless jokes about your patients and your colleagues. Well, you had your chance."

He presses a large red button on the tablet, a trapdoor opens beneath me and I find myself falling, falling, falling into blackness.

Suddenly I wake up, dripping with sweat.

Or not ...

Today when someone asks me what I do, I tell them that I'm now a social scientist. But I also say that the main reason I know a lot about life is that I used to work on the wild side. I can never forget—and I'm proud to say—that I was once a paramedic.

About the Author

After leaving the British Columbia Ambulance Service, Graeme Taylor took a master's degree in conflict resolution at Royal Roads University. His book, *Evolution's Edge: The Coming Collapse and Transformation of Our World,* won the 2009 IPPY Gold Medal for "the book most likely to save the planet." He married the woman who created the graphics for his book and they moved to Australia, where amid the palm trees and parrots he completed a PhD.

Graeme describes his present focus as "developing strategies for ensuring that our children will have a peaceful, prosperous and sustainable future. The underlying problem is that the global political economy is designed to support exponential growth, but constant growth is impossible on a finite planet. Constant growth is a cancer, a fatal disease. So I'm kind of a global cancer researcher. This critical work will keep me busy for the rest of my life."